CIRCULAR CITIES

With cities striving to meet Sustainable Development Goals, circular urban systems are gaining momentum, especially in Europe.

This research-based book defines the circular city and circular development. It explains the shift in focus from a purely economic concept, which promotes circular business models in cities, to one that explores a new approach to urban development. This approach offers huge opportunities and addresses important sustainability issues: resource consumption and waste; climate change; the health of urban populations; social inequalities; and the creation of sustainable urban economies. It examines the different approaches to circular development, drawing on research conducted in four European cities: Amsterdam, London, Paris and Stockholm. It explores different development pathways and levers for a circular urban transformation. It highlights the benefits of adopting a circular approach to development in cities, but acknowledges that these benefits are not shared equally across society. Finally, it focuses on the challenges to implementing circular development faced by urban actors.

This ground-breaking book will be essential reading for scholars, students, practitioners and policymakers interested in the circular economy, urban sustainability, urban ecology, urban planning, urban regeneration, urban resilience, adaptive cities and regenerative cities.

Jo Williams is the Director of the Circular Cities Hub. She is also an Associate Professor in Sustainable Urbanism at the Bartlett School of Planning, UCL. She has 23 years of research experience, working closely with industry, government and cities. She also advises International bodies (European Commission, EIB, EEA, OECD and UN) on the delivery of circular cities, low carbon cities, circular, sharing and bio-economies. Jo is on the steering panel for several large European research projects. She has published widely.

Routledge Studies in Sustainability

Sustainability Governance and Hierarchy
Philippe Hamman

Energy, Environmental and Economic Sustainability in East Asia
Policies and Institutional Reforms
Edited by Soocheol Lee, Hector Pollitt and Kiyoshi Fujikawa

Green Skills Research in South Africa
Models, Cases and Methods
Eureta Rosenberg, Presha Ramsarup and Heila Lotz-Sisitka

Sustainability and the Automobile Industry in Asia
Policy and Governance
Edited by Aki Suwa and Masahiko Iguchi

Smart Green World?
Making Digitalization Work for Sustainability
Steffen Lange and Tilman Santarius

Contesting Hydropower in the Brazilian Amazon
Ed Atkins

Circular Cities
A Revolution in Urban Sustainability
Jo Williams

For more information on this series, please visit: www.routledge.com/Routledge-Studies-in-Sustainability/book-series/RSSTY

CIRCULAR CITIES

A Revolution in Urban Sustainability

Jo Williams

Routledge
Taylor & Francis Group
LONDON AND NEW YORK

earthscan
from Routledge

First published 2021
by Routledge
2 Park Square, Milton Park, Abingdon, Oxon OX14 4RN

and by Routledge
52 Vanderbilt Avenue, New York, NY 10017

Routledge is an imprint of the Taylor & Francis Group, an informa business

Cover designed by Michael Lenz: Draught Vision Limited

British Library Cataloguing-in-Publication Data
A catalogue record for this book is available from the British Library

Library of Congress Cataloging-in-Publication Data
Names: Williams, Jo (Joanna), author.
Title: Circular cities : a revolution in urban sustainability / Jo Williams.
Description: Abingdon, Oxon ; New York, NY : Routledge, 2021. | Includes bibliographical references and index.
Identifiers: LCCN 2020046162 (print) | LCCN 2020046163 (ebook) | ISBN 9781138591141 (hardback) | ISBN 9780429490613 (ebook)
Subjects: LCSH: Sustainable urban development—Europe—Case studies. | City planning—Environmental aspects—Europe—Case studies. | Environmental policy—Europe—Case studies.
Classification: LCC HC240.9.E5 W55 2021 (print) | LCC HC240.9.E5 (ebook) | DDC 307.1/416094—dc23
LC record available at https://lccn.loc.gov/2020046162
LC ebook record available at https://lccn.loc.gov/2020046163

ISBN: 978-1-138-59114-1 (hbk)
ISBN: 978-0-367-74816-6 (pbk)
ISBN: 978-0-429-49061-3 (ebk)

Typeset in Bembo
by codeMantra

I dedicate this book to my courageous Father, without whose insight, knowledge and unerring belief in my ideas, it may never have been written.

Thank you Dad.

Anthony George Williams 9th March 1945 – 15th May 2017

CONTENTS

FIGURES

TABLES

PROLOGUE

Circular cities

Cities are urban ecosystems in which resources are consumed and waste (including emissions) is produced from a myriad of activities. Thus, as the urban population grows rapidly around the world, cities are increasingly responsible for global resource depletion and climate change. This could be addressed in part through a more efficient use of urban resources. By reusing and recycling land, infrastructure, water and materials or recovering energy from "waste", resource loops maybe closed (at least partially) within city-regions. Thus, cities become more resource-sufficient (consuming less from global stocks) and wasted resources are avoided. This approach also reduces the global transportation of resources, the use of fossil fuels and waste going to landfill, resulting in a reduction in greenhouse gas emissions. Thus, the global impact of cities is reduced.

Urban ecosystems are also degrading. Street trees are removed, culverts are filled in, gardens and parks are covered in concrete, waterways are choked with litter and land is contaminated by industrial uses. Thus, ecosystem services deteriorate. Natural cycles (water, carbon, nitrogen, potassium, soil formation) and biological processes (respiration and photosynthesis) are interrupted by human activities, resulting in pollution (air, water and noise), soil degradation, flooding, drought and urban heating. The loss of vegetation also reduces the potential for cities to sequester the carbon-dioxide they produce. This creates unhealthy and unstable living environments. Ecosystem services could be regenerated, natural cycles and biological processes restored through better management of the ecosystem (e.g. water, waste, conservation of habitats, horticulture, agriculture, forestry) and the inclusion of connected green and blue infrastructure into the urban fabric.

Urban ecosystems are constantly evolving, adapting to the context in which they are embedded. If cities cannot evolve, infrastructure becomes redundant, sites lie vacant, economic activities decline and social problems result. This can be seen in the shrinking cities of the USA, Japan, South Korea and Europe. However, in order to be able to adapt to change, urban infrastructure must be flexible, whilst communities and institutions must be able to learn and adjust to a "new normal". This will avoid a socio-technical lock-in and the associated waste of physical and human resources. By creating recyclable or adaptive urban environments we can avoid the waste generated by redundancy, vacancies and demolition. By enabling communities to learn from experiences, self-organise and be more engaged in the co-provision of infrastructure and services, we can also begin to their build capacity to adapt to change. By encouraging more temporary, pop-up activities which can adapt to context we also increase the resilience of cities.

Looping, ecological regeneration and adaptation enable the creation of circular resource flows; support natural cycles; and enable the city to renew (or recycle) itself. All three are circular processes. These three processes, operating together, underpin the theoretical conceptualisation of *Circular Cities* and *circular development.* All three processes reduce the impact of the urban population on the global ecosystem, whilst creating healthier local ecosystems in which the urban population can reside. This may seem a rather utopian vision. Yet cities in Europe are beginning to adopt circular principles and processes in their development decisions and policies. Circularity is also being incorporated into urban economic activities and the way in which services and infrastructure are designed. Urban inhabitants are being encouraged to embrace circular practices. Indeed, many European cities self-identify as *Circular Cities.* The European Union suggested that a circular approach will be adopted in cities to recover from COVID-19. This is supported by the European New Green Deal. Four such *Circular Cities* are presented in this book.

The research

For the benefit of an academic audience I will now briefly explain my approach to the research. This is a research-based book, which has drawn from the findings of three projects *Circular cities: strategies, challenges and knowledge gaps* funded by UCL Global Engagement Fund; *Circular Cities: London Circular Experiments* and *a Comparison of European Circular Cities* both funded by the UCL Sustainable Cities Grand Challenge. These three projects sought to determine the following:

1. The typology of circular development pathways
2. The benefits and disbenefits of adopting a circular development approach in cities
3. The challenges to circular development
4. The synergies and conflicts between circular actions and other urban strategies

Three sources of data were used to respond to the research aims 1–3. These were secondary data (academic papers, technical reports, policy documents and media reports), expert focus groups (which included policy-makers, urban technical professionals and service providers operating in cities across Europe) and detailed interviews with experts in case study cities, operating examples of best practice. The literature and interview responses were analysed using a combination of inductive and deductive content analysis.

The synergies and conflicts between circular actions and other urban strategies were explored using a systems approach. System maps, influence diagrams and causal loop diagrams were developed for the four case cities. Potential synergies and conflicts between circular actions and other strategies were initially hypothesised. The evidence (i.e. technical reports, interview data and academic papers) was interrogated to determine whether the predicted relationships manifested in practice. This enabled the author to determine how the circular actions influenced each other and how they were influenced by other urban strategies (i.e. localisation, optimisation, substitution and sharing).

This book not only presents a new theoretical development model but also clearly adds to the theoretical understanding of circular cities and circular development. It produces a typology of circular development pathways, identifies challenges to and levers for implementation, identifies the benefits of adopting circular development and explains the conflicts and synergies between circular development and other urban strategies. Thus, it provides a very broad and solid base from which others can begin to research this area. It does also highlight many questions which still need to be answered.

The book

The book is divided into three parts. The first part focusses on the conceptual development and underpinnings for circular cities and circular development. Chapter 1 outlines some of the ecological problems facing cities globally and the case for adopting a circular city approach. Chapter 2 presents the current conceptualisation of a circular city and suggests that a new conceptualisation is needed. Thus, it advocates a shift in focus from circular business models and industrial production processes in cities (an economic focus) to circular urban systems (a socio-ecological focus). It develops new conceptualisations for circular cities and circular development, which are explored throughout the rest of the book.

The second part of the book illustrates circular development principles operating in practice, using four detailed European case studies: Stockholm, London, Amsterdam and Paris. For each city the approach taken to circular economy and circular development is examined. Examples of different circular development pathways are presented. The levers for implementation, benefits of adopting the pathways and challenges to implementation are discussed. The four cities are extremely diverse in their approach to circular development.

Chapter 3 presents the case of Stockholm. Stockholm does not have a circular strategy. Nevertheless, circular thinking has been applied to the urban system and development for several decades. Thus, the development regime has already transformed and is supportive of the circular approach. Motivations for adopting circular principles are environmental. Stockholm demonstrates two distinct pathways for circular development. The first pathway is a strategic, city-regional approach based on the *ecocycles* (waste-to-energy) system. The second pathway uses planned eco-districts (Hammarby and Stockholm Royal Seaport) to demonstrate and test the application of the circular actions in a new build development. The case studies also demonstrate the dynamics between circular actions and other urban strategies.

Chapter 4 presents the case of London. London has a circular economy strategy, focussed on the creation of circular businesses and industrial sectors in the capital. Motivations for adopting this strategy are economic and environmental. The strategy does not conform with the definition of circular development provided in Chapter 2. Nevertheless, the London spatial plan addresses all three circular actions (albeit separately) and thus London does offer examples of circular development. The London cases demonstrate two distinct pathways. The first pathway uses a planned eco-district to demonstrate and test the application of the three circular actions in a new build development (Queen Elizabeth Olympic Park). The second pathway adopts a grass-root, temporary, experimental approach to delivering circular actions in an existing neighbourhood (Brixton).

Chapter 5 presents the case of Amsterdam. Amsterdam has a circular strategy, which is clearly linked to sustainable development. Motivations for adopting this strategy are economic and environmental. Amsterdam recognises the difference between circular economy and circular development. It has programmes in place to address both. It also demonstrates two distinct pathways for circular development. The first pathway encourages a strategic, city-regional approach to looping construction and organic waste (Circle City Scan). The second pathway adopts a grass-root, temporary, experimental approach to circular development (De Ceuvel).

Chapter 6 presents the case of Paris. Paris has a circular strategy, which offers a more holistic and integrated conceptualisation of circular economy, closer to this book's definition of circular development. It recognises the linkages between looping, ecologically regenerative and adaptive actions. It also territorialises these activities. Motivations for adopting this strategy are economic, environmental and social (solidarity). The social benefits are extremely important in this case. Paris demonstrates three pathways for circular development. The first pathway encourages a city-regional approach to looping construction materials, food and water. It also coordinates the strategic reuse of sites (e.g. Paris Reinvented). The second pathway adopts a grass-root, temporary-experimental approach to circular development (e.g. Les Grand Voisins, Bellastock), through the adaptive reuse of sites. The third pathway demonstrated by Clichy Batignolles uses a planned eco-district to demonstrate and test the application of the three circular actions in a new build development.

The third part of the book draws together the key findings from the four circular cities studied, answering the implementation questions how, why and what? Chapter 7 answers the "how" questions. How do circular cities and circular development pathways manifest? How are they implemented and what are the levers for transformation? How do the circular actions interact with each other and urban strategies? In Chapter 7, a typology of circular development begins to emerge. A range of levers (regulation, capacity building, provisioning powers) and resources (land and finance) which can be used to encourage circular development are identified. Finally, the synergies and conflicts between circular actions and other urban strategies are explored.

Chapter 8 discusses the reasons for adopting a circular development pathway, illustrated by the case studies. In addition to the environmental benefits, the pathway creates healthier, more adaptive living environments; a more diverse economic base with a range of jobs; greater community engagement and social solidarity. These benefits are not equally shared across the population, thus producing social inequalities. However, it is argued that the social problems highlighted could be addressed, given the political will to do so.

Chapter 9 explores the challenges to implementing circular development in cities. A range of challenges are presented. The case studies demonstrate that a shift in political philosophy, cultural values and assumptions underpinning the economic system will be required for the circular transformation of the existing development regime. However, the case studies demonstrate that where there is the political will and the resources, circular transformations can occur. However, circular experiments will need long-term, political and financial assistance and a supportive regulatory framework to ensure they scale-up.

The final chapter responds to the four questions set in Chapter 2. It defines circular cities and circular development. It discusses the dynamics between circular actions and other urban strategies. It highlights the levers for implementing circular transformations and the reasons for adopting circular development pathways in cities. It explores the challenges to circular urban transformation and suggests areas for future research.

Whether you are an academic searching for the definition of circular development, a student wanting to see the range of exciting circular projects emerging in cities across Europe or a practitioner/policy-maker wishing to understand the challenges to implementing circular development, this book provides a useful source of examples, guidance and inspiration. I hope you enjoy it.

ACKNOWLEDGEMENTS

I would like to thank UCL Sustainable Cities Grand Challenge and UCL Global Engagement Fund for funding the research on which this book is based. I would also like to thank UCL Sustainable Cities Grand Challenge for funding the establishment of the Circular Cities Hub (of which I am the Director).

I would like to thank Josephine Hintz (ICLEI) who worked with me on one of the research projects. I would like to thank those who kindly read and reviewed my draft. I would like to thank Michael Lenz for his front cover and graphics. Finally I would like to thank Sally Williams for her beautiful sketches of the circular projects.

PART I
Conceptualising circular cities

1

THE ECOLOGICAL CRISIS

The context

In a world facing resource scarcity, climate change, waste mountains and environmental degradation, it is time to rethink our role as custodians of the environment and resources. As centres of population (and often global affluence) cities consume the majority of the world's resources and produce the majority of its "waste". They are key contributors to the emission of greenhouse gases and climate change. Cities are also detrimentally affected by resource scarcity (and resource security issues) and the impacts of climate change.

Urban environments are fragile and often degraded by urban activities (e.g. motorised travel, shopping, manufacturing, construction, generation of energy). The degradation of ecosystem services in cities reduces the ability of the urban ecosystem to regenerate. This can result in air, land and water pollution; flooding or drought; urban heating and loss of productive soil and biodiversity. It creates unhealthy living environments, which are risky places to invest in and in the long-term are unsustainable. The regeneration of failing urban communities and ageing infrastructure is also integral to the health of the urban ecosystem. Attention to the natural and human aspects of the urban ecosystem is essential if the city is to be successful.

Cities are also in a constant state of flux. International demographic, climatic, technological, economic and cultural changes place different requirements on cities. The ageing population, mass migration, pandemics, climate change, the emergence of big data, globalisation, industrial restructuring and individualism have all had substantive impacts on our cities, not least on urban form and infrastructure. For example, globalisation has produced shrinking cities (in Europe and the USA), with dwindling populations and under-utilised infrastructure and land. The culture of individualism has boosted demand for private transport and accommodation for one-person households. Climate change requires the adaptation of cities to reduce problems of heating, flooding and drought (e.g. sponge cities in China). In all instances the adaptive capacity of the urban ecosystem (communities, infrastructure and urban form) to adjust to accommodate these changes is key to a city's longevity.

Resource crisis, waste and security

Currently cities consume 60–80% of natural resources globally. They produce 50% of global waste and 75% of green-house gas emissions (Camaren and Swilling, 2012). Large cities will account for 81% of total consumption and 91% of consumption growth between 2015 and 2030 (McKinsey, 2016). There is an imperative for cities to transition to increasingly circular economies to reduce the absolute magnitudes of global waste streams and emissions (Liang and Zhang, 2011). The UN estimates that 66% of the world's population will live in cities by 2050 (United Nations, 2014) while the global urban footprint will triple over the years to 2030 (Seto et al., 2012). There are three key drivers for this: increasing size of urban population, increasing affluence and greater distances over which goods (and materials) and waste travel. There are substantial accumulations of natural resources in buildings, infrastructure, products and waste deposits.

At a time when resources are becoming increasingly scarce, these technospheric resource reservoirs might offer an opportunity for more sustainable development, or at least provide a local alternative to imported, virgin materials and recycling of waste (Krook et al., 2012).

Until recently much of the municipal waste produced in European cities has been landfilled or exported to Asia (predominantly China and India) to be recycled (European Union, 2014). This has created waste dumps in Asia with serious environmental and health implications (particularly from e-waste). However, the introduction of more stringent regulations in some Asian countries (e.g. "operation green fence" in China, 2013) has improved practices. Unfortunately, it has also resulted in low grade municipal waste being exported elsewhere (e.g. Vietnam, Malaysia). More challenging markets, especially paper and plastics, have reduced Chinese interest in recycling, which creates a disposal problem for European cities. Meanwhile companies in Europe looking to recycle materials are starved of a supply (Laville and Taylor, 2017).

Urban resource security issues are increasingly a problem, particularly for water, food and energy. Currently, half of the world's cities with more than 100,000 inhabitants are situated in areas experiencing water scarcity (Richter et al., 2013) and the number of water-stressed cities is growing rapidly. The loss of agriculturally productive land surrounding cities is another key concern. Asian and African cities are experiencing food security issues (Brinkley et al., 2013). Rising land values caused by urban expansion has put pressure on farmers to either sell or convert to high-value activities. This reduces the hinterland's capacity to support urban demand for food. Meanwhile, cities have become increasingly reliant on global food producers, dramatically enlarging cities' resource hinterlands and exacerbating food security issues.

Cities consume 60% of global energy and are largely reliant on fossil fuels, which makes them particularly vulnerable to hikes in fuel price and energy embargoes (IEA, 2008). The generation of clean energy is severely hampered in cities by urban morphology (e.g. a lack of space, canyoning, overshadowing) and public opposition to energy generating plants. Renewable installations can often be found in the periphery, but urban demand is so high that it can be hard to supply all the energy requirement without reverting to fossil fuels (often imported) and nuclear power. Climate change exacerbates resource security problems by increasing the likelihood of more frequent natural disasters which affect provision.

Land and property speculation in *hedge cities* (e.g. London, Melbourne, Tokyo) has also resulted in vacant properties and sites (Cashmore, 2015; Sassen, 2015; United Nations, 2017). Global capital has been invested in land and housing as a commodity, as security for financial instruments that are traded on global markets and as a means of accumulating wealth. This financialisation of land and housing disconnects them from their social and environmental functions. Scarcity increases the value of land and properties. Thus, vacancies remain in markets where there would otherwise be oversupply (Cashmore, 2015). Vacancies in properties prevent reuse and result in the under-utilisation of the resource.

The degradation of the urban ecosystem

Ecosystem services[1] are integral for the long-term sustenance and renewal of the urban ecosystem, environmental regulation, as well as the health of the population (Demuzere et al., 2014; Gómez-Baggethun and Barton, 2013). The loss of these services is becoming increasingly important in stressed urban environments suffering from flooding, heating, pollution, declining biodiversity and soil degradation. These problems are also exacerbated by climate change.

This has implications for those living in cities. For example, large-scale floods displaced at least 100,000 people in over 1,800 cities in 40 countries (mostly in the developing world) from 2003 to 2008 (Kocornik-Mina et al., 2020), whilst 97% of cities in low- and middle-income countries (with more than 100,000 inhabitants) do not meet World Health Organisation's air quality guidelines (World Health Organisation, 2018). These problems are interdependent. For example, urban flooding, heating, pollution and declining biodiversity result in soil degradation. Urban flooding is exacerbated by soil compaction and sealing. Urban heating increases the effect of air pollution and reduces biodiversity. The degradation of urban ecosystem services has direct resource implications for energy, water and land. For example, urban heat islands increase the consumption of energy for air-conditioning in cities.

Increasing land values has resulted in urban densification and the loss of green and blue infrastructure providing ecosystem services (Bolund and Hunhammar,

1999; Gómez-Baggethun and Barton, 2013). This is particularly the case in *hedge cities* where in order to release maximum land value investors have applied to build high-value activities on the land (luxury residential or commercial space). Thus, land has been lost for lower value activities, for example, industrial activities and green space (Ferm and Jones, 2016). These activities are essential for the local production of resources; recycling and reuse of wasted resources; and regeneration of urban ecosystem services.

Failing communities and ageing infrastructure also threaten the health of the urban ecosystem. For example, in shrinking cities (e.g. Detroit, Leipzig) properties and land lie vacant. Between 2000 and 2016 there were 55 cities with declining populations, usually as a result of global economic or national demographic trends (United Nations, 2016). Population decline will occur in 17% of large cities in developed regions and 8% of cities globally from 2015 to 2025 (Woetzel et al., 2016). This will result in the under-utilisation of ageing infrastructure in cities, including water, sewage, transport, education and health systems, leading to wastage of resources (Rink et al., 2012). Thus, scale appropriate systems will be needed to support smaller populations. Vacant land can provide opportunities for urban transformation (Németh and Langhorst, 2014). It may be possible to encourage the integration of new industrial uses (enabling industrial symbiosis) and blue and green infrastructure (to protect ecosystem services) as part of the regeneration process.

Adaptive capacity and resilience to long-term change

Cities must maintain an optimal "fit" with their dynamic environment (Rauws and de Roo, 2016) if they are to avoid the wastage of resources, preserve the urban ecosystem and remain viable. For example, cities will need to adapt to climate change, to ameliorate problems caused by flooding, heating, pollution, drought and so on. Shrinking cities will need to tackle the under-utilisation of infrastructure and wasted land. This resilience to change is built through adaptive capacity of both the urban community and infrastructure (the socio-technical system). However, the coevolution of the socio-technical system may be limited by current social practices and lifestyles of those living in a city; its physical form, infrastructural systems and local environment; institutional inertia to change; regulation and economic cost.

The adaptive capacity of a city is underpinned by the potential for the urban community to self-organise and learn and for the socio-technical systems to co-evolve. Self-organisation is a key property through which systems self-innovate and self-stabilise in response to changing circumstances (Rauws and de Roo, 2016). Informal and formal networks for learning are also essential for systemic transformation. Designing infrastructural, spatial and institutional frameworks which allow greater flexibility and incremental change is essential for building adaptive capacity (Teisman and Gerrits, 2014). However, those living in cities may lack the knowledge, skills, social networks or financial resources to enable them to easily self-organise and adapt to change. This is particularly problematic for low-income, poorly educated groups. Thus, building adaptive capacity in cities should be supported by government.

So what is needed?

To begin to address the challenges facing cities in the twenty-first century we need to ensure they are **resource efficient, ecologically regenerative and resilient**. This will enable cities to address directly three key sustainability challenges often overlooked: futurity, inter-generational equity and environmental protection. Indirectly adopting such an approach should also help to address the health and well-being of those living in cities and the creation of sustainable urban economies. In theory, taking a circular approach to development will enable the resource-efficient, waste-free, ecologically regenerative and continual renewal of the city.

Note

1 Eco-system services support nutrient cycling, soil production and flood control. They can produce resources (e.g. energy and food) and regulate urban systems (e.g. carbon sequestration, climate regulation, and air and water purification).

2

MOVING FROM A CIRCULAR ECONOMY TO A CIRCULAR CITY

Circularity derives from an ecological conceptualisation of the world. The focus shifts from linear systems, which consume an infinite supply of new resources (inputs) and produce "waste" (outputs), towards circular systems, in which resources are reused, recycled or recovered. The principle of circularity has been applied to industrial systems (industrial symbiosis), production processes (cradle-to-cradle) and economic systems (circular economy).

Industrialists developed the idea of industrial metabolism in the nineteenth century. Industries metabolised resources, producing outputs – often classified as waste – which could be used by other industries (Simmonds, 1862). By 1930, industrial symbiosis had appeared in the literature (Fischer-Kowalski and Haberl, 1998; Parkins, 1930). Industries formed symbiotic relationships with each other enabling by-products (water, energy and materials) from one industrial process to be used by another. This was often facilitated by close physical proximity.

The ideas of industrial symbiosis were further developed in the 1990s by industrial ecologists, who viewed industrial systems as an ecosystem (Chertow, 2007). This has led to the creation of examples of industrial symbiosis operating at a national level in eco-industrial networks (e.g. NISP in the UK). The tendency to date has been to focus on industrial networks or industrial parks. More recently there has been some mention of the application of industrial symbiosis to cities. Arguably this is not the application of circularity or symbiotic principles to the urban system, but more about creating symbiotic relationships between industries which happen to operate within an urban system.

Cradle-to-cradle thinking has been applied to the production of goods through the design and manufacturing processes. It aims to reduce waste produced during the lifecycle of a product, through minimisation, reuse, recycling and recovery of resources. The current model is based on a system of "lifecycle development" initiated in the 1990s (McDonough, 2002). Life cycle assessment compares the full range of environmental impacts from products and services, throughout their lifecycle (from extraction to disposal). This is achieved by quantifying all inputs and outputs of material flows and assessing how these material flows affect the environment. This conceptualisation has been applied to buildings and building components, but not to cities.

Circular economy

The closed-loop (circular) economy first emerged in the sixties (Boulding, 1966). It was this conceptualisation which became influential upon German and Japanese economic policy during the 1980s and 1990s (Bilitewski, 2007; Moriguchi, 2007) and encouraged the adoption of circular principles in business and industry. Circular economy (CE) is a model for production and consumption (with an emphasis on production), whose ultimate goal is to achieve the decoupling of economic growth from natural resource depletion and environmental degradation (Jackson, 2009). The focus of CE is of course on the economic system, usually sectors (e.g. food, construction, electronic goods), or business models for specific companies (CE100). However, none of the three conceptualisations – industrial symbiosis, cradle-to-cradle, CE – have focussed on the urban system as the bounded area of analysis.

CE places emphasis on the redesign of processes and cycling of materials within the economic system. It aims to "design out" waste, return nutrients and recycle durables using renewable energy to power the economy (UNEP, 2006). Thus, CE is not merely seen as a preventative approach, but as an ecologically restorative and regenerative approach, repairing previous damage by designing better economic systems (EMF et al., 2015; UNEP, 2006).

The Ellen MacArthur Foundation (EMF) developed the RESOLVE framework for CE (Figure 2.1). It defines CE as one that provides multiple value-creation mechanisms which are decoupled from the consumption of finite resources. It describes six actions which are critical to the transition to a CE:

1. Ecological regeneration through a shift to renewable energy and materials, alongside the return of recovered biological resources to the biosphere
2. Keeping components and materials in closed loops (reuse, recycle, recover, remanufacture), prioritising inner loops (e.g. reuse) and thus reducing waste
3. Sharing resources to keep product loop speed low and maximise utilisation of products to reduce waste
4. Optimisation of the performance and efficiency of products, alongside the removal of waste in production and supply chains, leveraged by big data
5. Dematerialise resource use by delivering utility virtually
6. Replace existing products and services with lower resource consuming options.

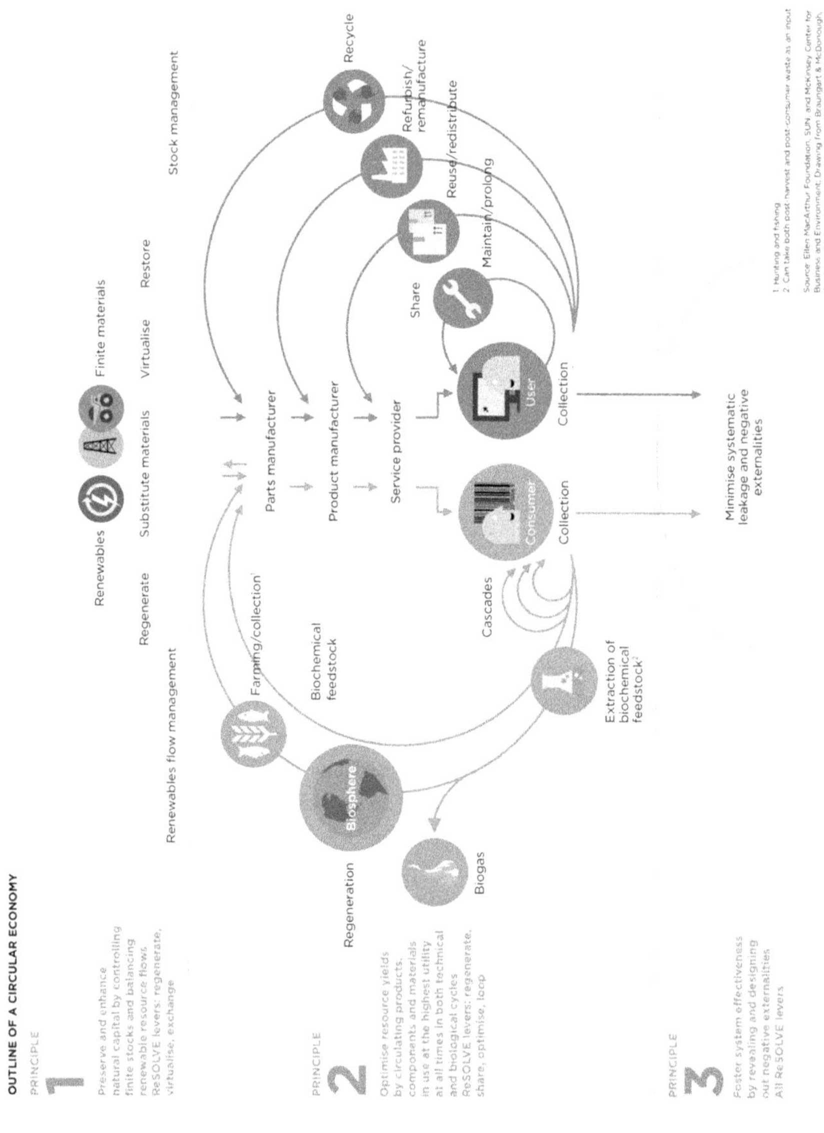

FIGURE 2.1 RESOLVE – framework for a circular economy.
Source: EMF et al. (2015).

RESOLVE is the most widely used CE framework for businesses, partly as a result of its promotion through the CE100 network and inclusion in a *Circular Economy Vision for a Competitive Europe* (EMF et al., 2015). An analysis of existing circular business model types (26 in total) demonstrated that RESOLVE provided the most comprehensive framework for moving towards a CE (Lewandowski, 2016). RESOLVE is indeed useful for conceptualising a CE. However, it is less appropriate when conceptualising a circular city and circular development (Williams, 2019a).

Moving from an economic system to an urban system

RESOLVE is designed to produce circular practices in an economic system, particularly within businesses or industrial sectors. The focus is largely on increasing economic efficiency within production systems which result in environmental benefits. The economy is pre-eminent and cuts across spatial boundaries. It is governed at a national or international level. The goals are largely economic and relate to the accumulation of capital and wealth. The focus is on businesses or industries operating in cities, rather than on systems of provision (services and infrastructure).

In contrast, a circular urban system – a circular city – is spatially bounded. It is governed at a local level, although local decisions and processes are deeply affected by international and national regulatory and economic systems. Nevertheless, local actors, particularly local government, are focussed on delivering a range of societal benefits. The economy and economic goals are not pre-eminent. The focus is also on systems of provision – infrastructure and services – rather than on systems of production. Thus, a circular city is distinctly different from a circular economic system.

RESOLVE focuses on production systems. However, cities are more often centres of consumption, not centres of production. Many resources consumed by urban inhabitants are produced outside cities (especially in the European context). The shift in emphasis in cities should be towards resources consumed and subsequently "wasted" by its inhabitants. Thus, we need to shift focus from systems of production in an economic system to systems of provision in an urban system. RESOLVE focuses on small-scale systems of production, usually within organisations or a single industrial sector. It does not consider complex urban systems of provision, across multiple sectors. Nor does it consider how these systems of provision interact with the varied lifestyles of those living in cities, producing different social practices. Furthermore, RESOLVE does not focus on lifestyles, and how lifestyles themselves can influence citizens' willingness to adopt circular practices. Thus, in a conceptualisation of a circular city, an emphasis should be placed on consumption (delivering circular lifestyles and practices) and systems of provision.

RESOLVE provides no indication of where the technological and biological processes integral to a CE take place. Thus, it does not conceptualise a spatial dimension. Yet cities are a physical entity, anchored in a specific location. The spatial dimensions of a city are based on physical form (land-use) and a variety of functions (e.g. commute patterns, water catchment area). Thus, it is important in the conceptualisation of a circular city to determine where resource flows, waste assimilation and circular actions will happen. It is also important to determine at which scale closing resource loops would be most appropriate (neighbourhood, city or city-region).

RESOLVE focuses on technological and biological resources, but land and infrastructure are not considered. Yet these are important resources and should be included in a circular city conceptualisation. Land is a scarce and valuable resource in cities. Land use affects the feasibility of adopting circular actions (e.g. localisation of activities or looping of resources) within a city. It provides space for circular activities, which affect the city's ability to assimilate "waste" (Pandis, 2014). Land offers ecosystem services which are crucial for the regeneration of the urban ecosystem (Folke et al., 1997). Land-use patterns also affect urban activities and thus resource consumption and production. This influences resource flows and the capacity for loops to be closed.

RESOLVE also ignores infrastructure in its conceptualisation. Yet, infrastructure is a resource mine. Thus, the reuse or recycling of infrastructure must be prioritised in a circular city. Infrastructure also governs how resources are supplied, managed and consumed in cities (Chester and Allenby, 2018). It therefore influences the systems of provision and consumption of resources. Thus, infrastructure is critical to the delivery of a circular city.

Moving from an economic to an ecological focus

For circular cities and circular development, we shift from an economic to an ecological focus. Both the urban economy and society operate within the environmental carrying capacity of the urban ecosystem. In an ecological framing, resources are not infinite (as implied in neoclassical economic models) and global environmental carrying capacity limits the growth of cities. Thus, cities must operate within their ecological carrying capacity, if society and economy are to flourish. Operating within the ecological carrying capacity is a fundamental aim of circular cities.

Ecological footprint

An urban ecosystem will have an ecological footprint (appropriated carrying capacity). This is the area of biologically productive land and water required to produce the goods and services consumed in urban activities and to assimilate the wastes generated by the city's population. The urban ecological footprint extends beyond the physical or administrative boundaries of the city, but the aim for a circular city would be to reduce it. In order to limit the ecological footprint of a city, we must reduce the resources consumed (particularly finite resources) and waste produced by urban activities. For example, this could be achieved through resource looping and infrastructural adaptation. It is also important to increase assimilation of waste and production of resources locally. This can be achieved by increasing resource looping and ecologically regenerating the urban ecosystem.

Sufficiency

Urban ecologists highlight the importance of self-sufficiency for ecological optimisation in cities (Rosales, 2017). Self-sufficiency will also increase the resilience of cities to resource shocks. Sufficiency is achieved by staying within local and regional carrying capacity, by regulating patterns of consumption and restoring resources. The carrying capacity of the urban ecosystem and its ability to be self-sufficient is affected by the health of a city's ecosystem services (Rosales, 2016). The ecological regeneration of an urban system will increase its capacity to be self-sufficient.

Sufficiency also helps reconnect those living in cities with the environmental consequences of their consumption decisions, lifestyles and social practices. It ensures that both positive and negative externalities of resource consumption are localised (Rosales, 2016). Thus, it can help to drive the changes in lifestyles and social practices needed for urban populations to live within the ecological carrying capacity. However, moving towards greater sufficiency requires a more integrative approach to cities and their regions (Mumford, 1968). Thus, a city-regional approach is required.

These ideas are echoed by the economic concept of eco-localism. Localisation of resource production and the local benefits accrued from ecosystem services have beneficial social, environmental and economic outcomes (Curtis, 2003). Local symbiotic capital reinforces the preservation and restoration of natural capital, the functioning of a sustainable local economy and localisation of resource flows (Curtis, 2003). Local symbiotic capital is also fundamental to industrial symbiosis, which in turn can enable resource looping.

Closing resource loops

Urban ecologists describe cities as complex organisms which metabolise resources (Kennedy et al., 2007; Wolman, 1965). They are composed of a network of inter-dependent actors (producers and consumers) between whom resources flow (usually materials, water and energy). Cities are open systems and resource flows are usually linear. Thus, resources flow across administrative and physical boundaries and are lost from the urban system. By creating closed-systems, in which resources are looped (reused, recycled or recovered), waste generated by urban processes can be reduced.

Closed-loop systems can help to deliver decarbonisation and dematerialisation. Resource looping both at local and at global scales is seen as essential for reducing waste and emissions and improving the health of the urban ecosystem and global environment (Orr, 1992). Thus, closing resource loops is fundamental to a circular city. Despite the experientially slow adoption by cities (Kennedy et al., 2011), circular metabolism applications are gathering increasing momentum in planning thought (Agudelo-Vera et al., 2011). "Loop closing" was recently identified as one of the four dominant urban development types of post-networked pathways to low carbon futures for cities (Coutard and Rutherford, 2011).

Regenerative capacity

Urban ecologists also recognise the importance of a city's regenerative capacity, to produce useful biological resources and absorb waste generated by human activities. This is enabled by ecosystem services. However, ecosystem services in cities are often degraded. Many cities are in biological deficit, due to increasing demand being placed on resources, and rely on increasingly large hinterlands to sustain them. Allocation of land in cities for ecosystem services for production, to tackle the degradation of natural capital and environmental hazards, could potentially help to reduce the resource hinterland and regenerate the urban ecosystem. This would also help cities to operate within the ecological carrying capacity.

Adaptive capacity

Cities are dynamic and complex adaptive ecosystems, constantly evolving with a changing context (Geddes, 1915; Gunderson, 2000). Like all ecosystems, cities have the capacity to cope with disturbance and stress, returning to a stable state. This is influenced by the capacity of urban institutions, communities and networks to learn and store knowledge and experience. It is underpinned by creative flexibility and an inclusive approach to decision-making and problem-solving within the city. It is reinforced by the urban population's ability to self-organise to respond to challenges in the environment.

The adaptive capacity of physical form (i.e. urban form and infrastructure) is also key. Socio-technical lock-in to existing infrastructure and land-use patterns in cities often prevents adaptation. This becomes a problem when societal demands change and new systems of provision are needed. These changes can render infrastructure and spaces obsolete or at best under-utilised. This wastes resources in cities. Yet the demolition and renewal of infrastructure also has resource implications. To limit waste within the built environment, we need to plan for change and create some flexibility to enable the adaptation of urban form and infrastructure.

The importance of context

Urban ecologists recognise the interdependency between cities and their local environment. Context effects the carrying capacity of the urban ecosystem. It influences activities producing and consuming resources, generating and assimilating waste. Equally, local political priorities (and policies), regulation, economy, culture, social practices and so on can affect the ability of the urban ecosystem to close resource loops, be self-sufficient, to regenerate and to adapt. Thus, the capacity for a city to reduce resource consumption and waste and to go-circular will depend on the local context.

Defining circular cities and circular development

If we draw together these two lines of thought, we arrive at a socio-ecological conceptualisation of a circular city and circular development. A circular city is a socio-ecological system, consisting of a bio-geo-physical unit and its associated social actors and institutions. It is a complex, regenerative and adaptive system, delimited by spatial and functional boundaries, surrounding an ecosystem. There are three actions fundamental to both a circular city and circular development (Figure 2.2):

I. Looping actions (reuse, recycling and energy recovery) – a circular city is an open system with many linear processes; however, where possible these processes will be closed. This reduces waste and promotes the most efficient use of resources. Examples include waste-heat recovery systems; food-reuse cafes; bio-refineries, grey-water recycling systems; adaptive reuse of buildings and land reclamation.
II. Ecologically regenerative actions – regenerate the urban ecosystem and ecosystem services. Ecologically regenerative actions are often operationalised through the inclusion of green and blue infrastructure (e.g. permeable surfaces, reed-beds, retention ponds, green roofs) into the urban fabric or the management of urban ecosystems (e.g. conservation, farming, forestry).
III. Adaptive actions – build capacity within the urban fabric and communities to adapt to change. Capacity is built through the use of flexible design, collaborative planning, co-provision and systems for learning.

Combining all three actions will enable an urban system to renew itself, whilst minimising resource consumption and waste production. This enables the circular city to operate within its ecological carrying capacity.

Circular development is the process which integrates circular actions into urban systems of provision. It produces circular systems (e.g. grey-water recycling systems), circular activities (e.g. industrial symbiosis) and circular infrastructure (e.g. adaptable buildings). It can be driven by spatial planning or the economic development processes. Circular development closes resource loops at a variety of scales (i.e. sub-regional, regional, national, international), resulting in greater sufficiency at a city-regional level. It enables the reconnection of people with nature and development of circular practices, through urban form and systems of provision. Circular development also protects and enhances ecosystem services. This helps to reduce or assimilate waste and produce new resources within the urban system. Circular development creates adaptable cities, offering space for transformation and growth, as well as infrastructure which evolves with changing needs. The circular development process also enables learning within communities and encourages self-organisation.

A variety of circular development pathways are likely to emerge from different urban contexts, resulting from diverse political, economic, cultural, social, environmental, regulatory and technical conditions. Existing urban strategies will also influence the circular development pathway. These include strategies for optimisation,[1] substitution,[2] localisation and sharing[3] (Figure 2.2). It is important to understand the dynamics between circular actions and existing urban

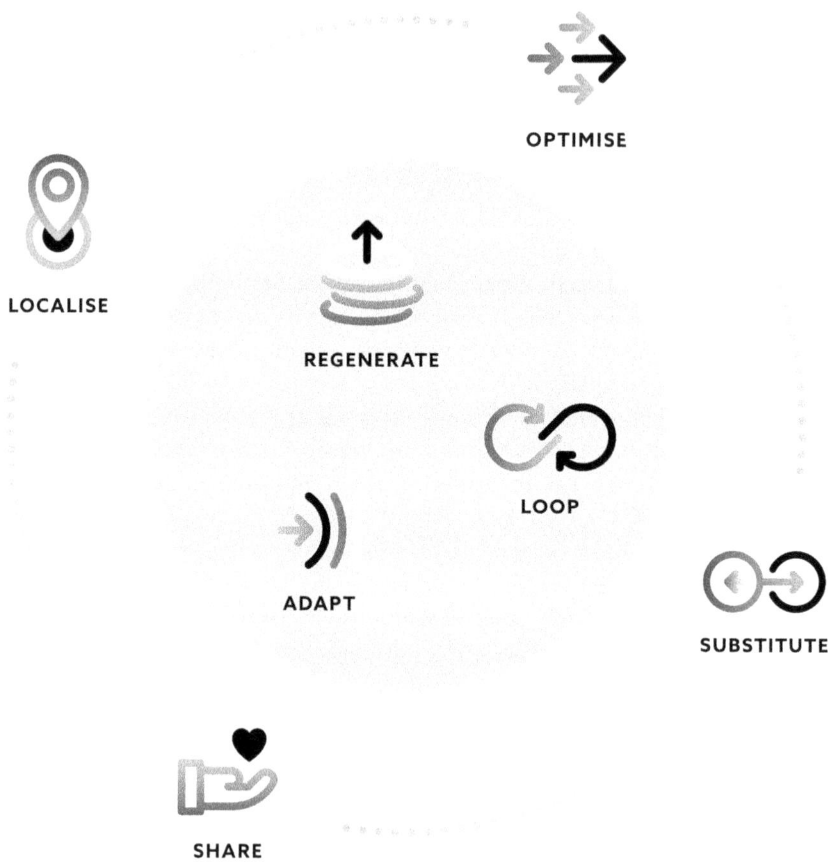

FIGURE 2.2 Circular development pathways – interactions between circular actions and existing urban strategies.
Source: Authors own produced by Draught Vision Ltd.

strategies, in order to understand the best approach to circular development within different contexts.

Circular cities: a European phenomenon

Post-2014, circular cities began to emerge in Europe. This followed the publication of a series of policy documents produced by the European Commission.[4] By 2016, the Netherlands, Scotland, Finland and Germany had national strategies for CE in place. By 2018, France, Slovenia, Portugal, Greece, Italy and Luxembourg had joined them.

The New Urban Agenda for Europe established a specific partnership for implementing CE at the urban scale (Partnership for Circular Economy, 2017). Initially, London, Paris and Amsterdam were the first cities to declare their intention to adopt a circular approach to urban economic development. They were soon followed by Peterborough, Copenhagen, Rotterdam, Glasgow, The Hague, Maribor, Almere, Birmingham, Brussels Capital region, Dusseldorf, Genoa, Ghent, Ljubljana, the Lyon metropolitan region, Munich, Oslo, Strasbourg, Turin, Rome, Marseille, Porto and Utrecht. More recently Lisbon and Berlin have begun to develop their circular strategies.

There is a great deal of variation in how these cities define "circularity". Some focus on the application of CE principles in the city-region. This may include all six actions presented in the RESOLVE framework, applied in different combinations. Other cities focus on supporting circular business models (London), encouraging industrial symbiosis (Rotterdam) or managing municipal material waste (Lisbon). In contrast, some cities have adopted a more holistic and territorial definition (e.g. Paris), closer to the concept of circular development presented in this book. Some recognise the importance of implementing the CE strategy as part of an integrated approach to sustainable development, alongside policies for ecological regeneration, adaptation and resilience (e.g. Amsterdam). Others have adopted a circular development pathway, but do not have an official circular strategy (e.g. Stockholm).

Motivations for adopting the circular development in cities also vary significantly. These motivations may include city-marketing and export of urban innovation (e.g. Amsterdam, London and Stockholm); social solidarity and redistribution of resources (e.g. Paris, Lisbon and Berlin); business development and job creation (London, Amsterdam, Paris); regenerating the local industrial base (e.g. Paris); resource security (e.g. Amsterdam, Lisbon, Paris) and tackling climate change (Paris, Berlin, Amsterdam, Lisbon and Stockholm). This variety of motivations will affect the ways in which the circular development pathways manifest.

In the next four chapters we visit four of these cities – Stockholm, London, Amsterdam and Paris – to observe how circular development has manifested in practice. This provides the evidence base for the rest of the book which seeks to:

- define circular development – common characteristics, variation and typologies;
- determine the dynamics between circular actions and other urban strategies;
- examine levers for implementing circular development;
- identify the reasons for adopting a circular development approach in cities; and
- finally to explore the challenges to circular urban transformation.

Notes

1 Examples of actions producing resource optimisation: introduction of smart grid, energy efficient buildings and vehicles, mass transit systems and community heating systems.

2 Examples of substitution: finite resources can be substituted with renewable resources (e.g. renewable energy); resource-based activities substituted with service-based activities (e.g. buying clean water rather than waste-water systems); activities requiring movement with virtual activities (e.g. teleworking); durable infrastructure substituted with non-durable infrastructure.
3 Resources can be shared in cities across a range of activities, including living (e.g. co-housing, library of things), working (e.g. co-working spaces) and travel (e.g. public transport and vehicle sharing schemes).
4 European Commission (2014, 2015).

PART II

Circular cities

European case studies

3

CIRCULAR STOCKHOLM

Circular Sweden

Sweden does not have a direct policy on circular economy. Yet it is a rich country and consumption levels are extremely high. The average citizen consumes 7.3 global hectares per capita which would require the equivalent of four Earths to sustain them (World Wide Fund, 2016). Imports of goods are high, especially from Asia. However, the population of Sweden is relatively small and the country is resource rich. Domestically Sweden has also decarbonised its energy supply through the use of nuclear and renewable energy (particularly hydropower and biomass) and energy recovery from waste. However, it is dependent on the importation of waste to sustain the energy recovery system.

Sweden has been a pioneer in environmental sustainability for many decades. It has prioritised ecological regeneration, resource conservation, building adaptive capacity and mitigating climate change. Hence, the protection and regeneration of urban ecosystems is well supported. The Swedish government has applied systems thinking and holistic solutions to policy-making and systems management. Circularity, the need to close resource loops and increase positive feedback has emerged from this. By 1995, governmental interest in the "natural step" and subsequent adoption of the Alborg Charter (Conference Sustainable Cities and Towns, 1994) embedded thinking which addressed looping, ecological regeneration and adaptation in Swedish cities (Williams, 2016).

The materialisation of circular thinking was encouraged through the allocation of funding streams[1] for new urban quarters. This funding enabled the development regime to implement circular systems in urban districts (Williams, 2016). It manifested as *ecocycles*, implemented in three urban districts: Malmo (B001) and Stockholm (Hammarby Sjöstad and Stockholm Royal Seaport). Ecocycles is fundamentally an energy recovery system. It converts under-utilised materials (sewage, domestic waste) to energy and reuses heat (e.g. from industrial processes, water-cleansing processes) for cooking, space-heating, appliances and transportation. Resource loops are closed by connecting the waste, energy, water and transport systems.

The technical capacity to deliver *ecocycles* developed in Swedish cities over several decades (Williams, 2016). District heating systems powered by waste, renewable energy and energy efficient buildings have been part of mainstream development models in Swedish cities for some time (Figure 3.1). The district heating system was introduced into Swedish cities during the 1950s to tackle pollution and increase energy efficiency. Originally, district heating systems were powered by fossil fuels. However, over a 50-year period, gas was substituted with waste, thus closing the resource loop and reducing greenhouse gas emissions. Initially, this transformation was driven by the oil and energy crises of the 1970s and the Municipal Energy Act 1977. Latterly it was driven by the European landfill tax (1999/31/EC) which required a reduction in waste materials being disposed of in landfill. Today, more than 90% of Swedish municipal waste is diverted from landfill and used to generate energy.

Circular development in Stockholm

Stockholm provides an interesting case for this book. It does not have a circular economy strategy (like Paris, Amsterdam or London), yet circular-thinking has been embedded into its policies, infrastructure and services for 25 years. This circular development pathway is reflected in Stockholm's spatial plan, sustainable development strategy and climate action plan. Circular principles first manifested in Stockholm (as Ecocycles I.0) in Hammarby. Hammarby developed the infrastructure required to create a closed-loop, waste-to-energy system.

More recently, circular principles have manifested in the circular regeneration of Stockholm Royal Seaport. In addition to the closed-loop system in the living environment, this development show cased closed-loop systems for the port (as Ecocycles 2.0). It also demonstrated ecological regeneration in its restoration of the waterways, caverns, soils and expansion of green infrastructure. It has implemented climate adapted environments and adaptive capacity has been built within communities using arrange of communication and engagement methods. SRSP tells us something about the dynamic between circular actions and other urban strategies. Both cases also provide us with greater understanding of the challenges faced in implementing a circular development pathway.

Circular Hammarby

Stockholm city council developed the eco-cycles system in Hammarby, with the aim of creating circular resource flows and improving resource efficiency (Figure 3.2). The system utilised the existing, proven city-wide infrastructural systems (city-wide district heating system, the Högdalen combined heat and

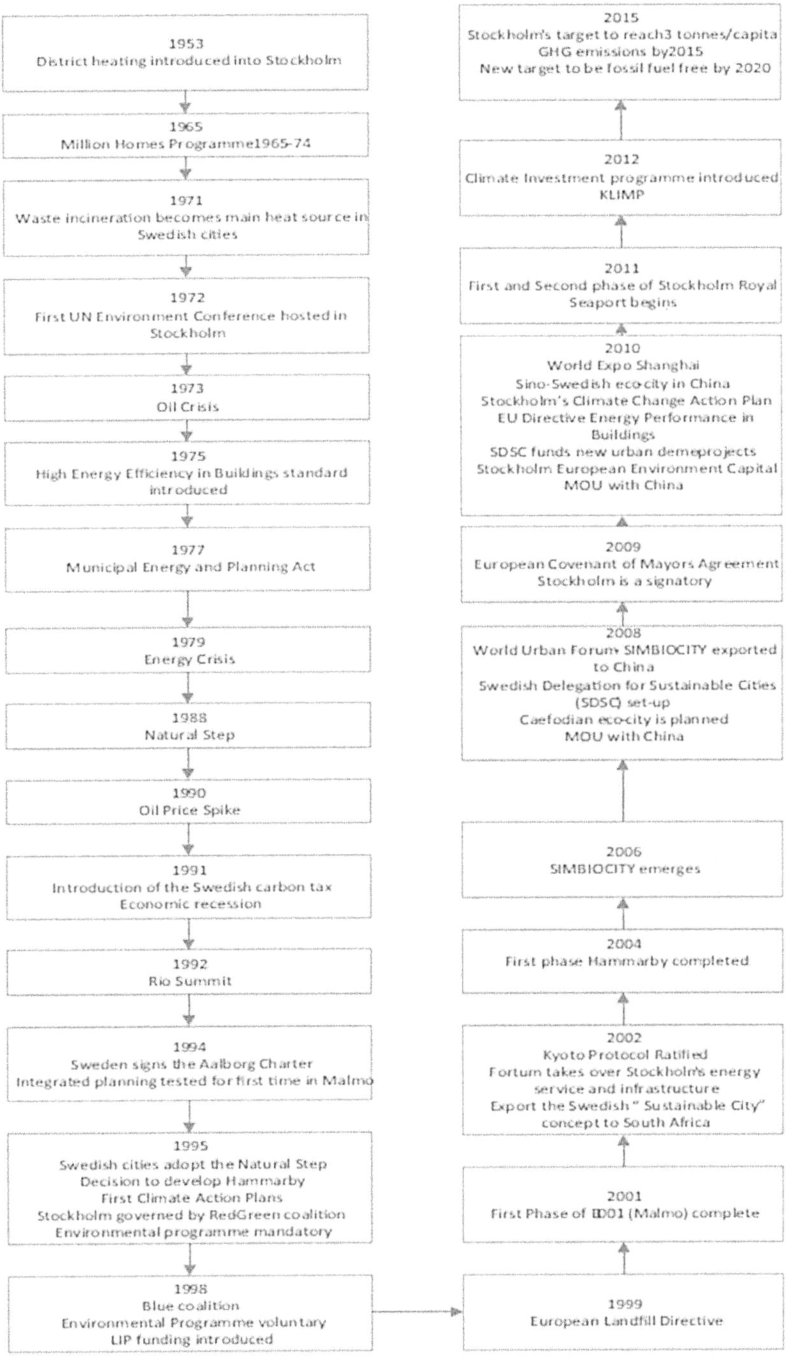

FIGURE 3.1 Historical time-line for the ecocycles system.
Source: Williams (2016).

power plant and the Hammarby thermal power station) together with new technologies for converting sludge into fertiliser and biogas. The heat produced from the process of purifying waste-water is used by the thermal power station. The buildings in Hammarby have also been designed to be more energy efficient (consuming 60 kWh/m^2/year) and to produce renewable energy on site using solar cells, solar collectors and fuel cells (Pandis et al., 2013). Of these innovations the biogas element has been most successful, used by buses in Stockholm and biogas cookers in Hammarby.

The ecocycles system has reduced non-renewable energy use by 28–42%; CO_2 emissions by 29–37%; water consumption by 41–46% and waste going to landfill by 90% (Brick, 2008). It demonstrates one approach to delivering circular resource flows at a neighbourhood or potentially city-scale. This has a beneficial impact on green-house gas emissions produced, by avoiding fossil fuels (used in heating and vehicles) and the production of methane from landfill. However, it is at the lower end of the circular hierarchy, as its focus is on energy recovery rather than on resource recycling or reuse. It also requires waste importation, which has implications for greenhouse gas emissions.

The ecocycles system was implemented in Hammarby through a coordinated action across several local government departments (planning, energy, waste, water and transport). Initially, the services and infrastructure integral to the ecocycles system were publically operated (at city or county level). This helped with infrastructural integration and goal alignment between stakeholders involved in the implementation of the system (Williams, 2019b). It also utilised existing infrastructure. This helped to avoid barriers created by sunk costs. The city coordinated the integration of resource streams between urban sub-systems, focussed on service delivery (i.e. providing affordable warmth; clean and accessible public transport; reducing waste going to landfill) rather than maximising unit throughput (Williams, 2019b). This service-based approach encouraged a more efficient use of resources (via exchange, recycling and recovery).

Nevertheless, some political pressure was needed to implement ecocycles and overcome the initial inertia within government departments. This inertia resulted from perceived high transaction costs including sunk investment costs, separate and parallel delivery of services, loss of control over systems difficulties communicating and negotiating systems integration (Williams, 2019b). The goals for effective service delivery across the city and county councils were largely aligned. Thus, institutional barriers diminished over time. Trust and understanding was built between actors which enabled effective management of the system. Thus, capacity to deliver circular resource flows developed within the city (ibid).

The planning process and the strategic plan were used as vehicles for implementing the system and ensuring its longevity. The collaborative planning process was used to engage and build support for the system amongst the service providers and developers. The strategic plan guaranteed that both urban form

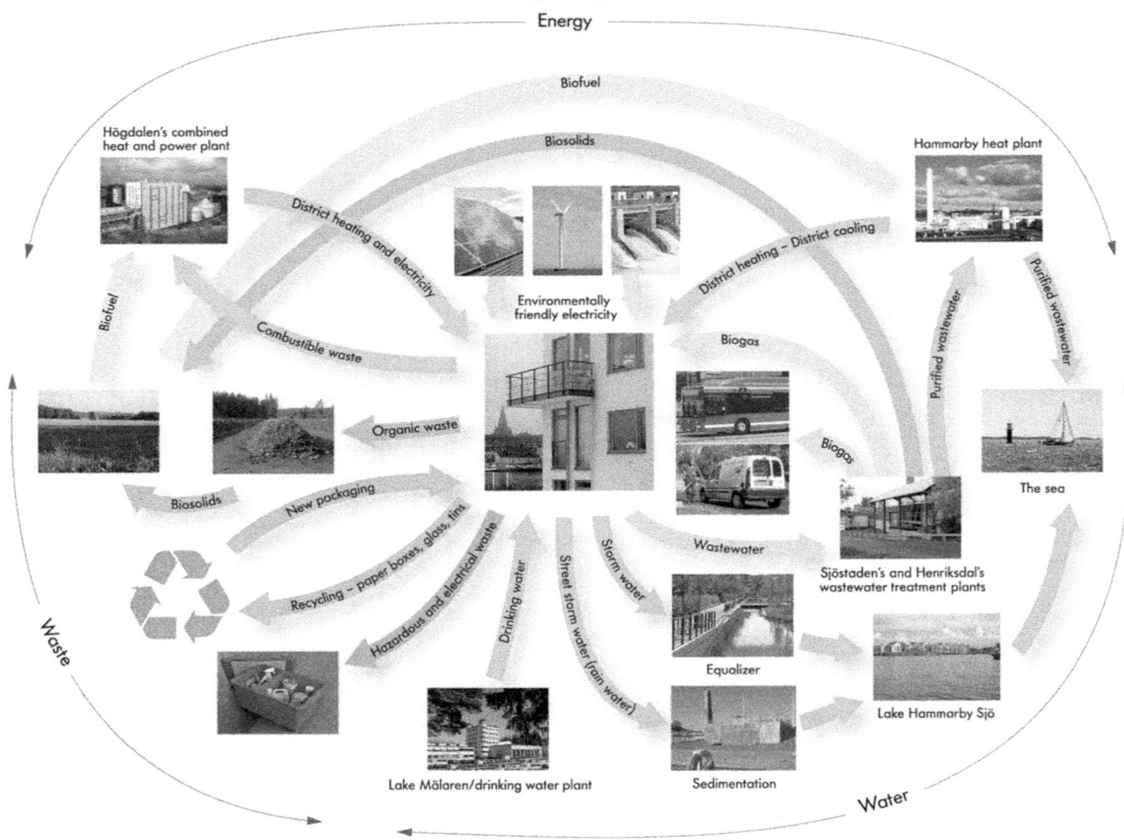

FIGURE 3.2 Ecocycles system.
Source: Stockholm City Council Website. http://www.hammarbysjostad.se/ (accessed 24-04-20).

and the development of new infrastructure would continue to support the expansion of the system across the city (ibid).

There was a lack of involvement of residents in the design process for Hammarby. After the post-construction evaluation, the Stockholm City Council realised their mistake of not including the community in the process. However, this was hard to do given it was a speculative development. The GlasHut (an information centre for residents) was set up to provide environmental education programmes and technical advice for residents and businesses operating on site. However, success in changing social practices was limited. Evidence for this could be seen from the misuse of the vacuum waste system and increased expenditure on energy hungry activities outside the home. Although residents were attracted by the environmental profile of the Hammarby development, they remained passive consumers with limited pro-environmental behaviour. Ecocycles I.0 could have produced greater resource savings, had the community been more engaged in the design and development process.

Circular Stockholm Royal Seaport

Stockholm Royal Seaport (SRSP) is the second generation model for circular urban development in the city. It combines all three circular actions in its development and operational processes. In the Royal Seaport, the development corporation aimed to move up the waste and energy hierarchies, to encourage more recycling and reuse of materials and use of renewable energy. Processes were established in SRSP, to build a greater degree of civic engagement, in the design and operation of the development, in order to change social practices.

SRSP is an important port hub for the movement of both freight and people. Annually 9.7 million tonnes of freight and 16 million passengers pass through the port (Ports of Stockholm, 2018). It is Sweden's third largest freight port and number one passenger port. It creates 8,000 jobs in the region (Ports of Stockholm, 2018). SRSP covers an area of 236 hectares (City of Stockholm, 2017). The land is owned by the City of Stockholm and is very close to the city centre (3.5 km). It is well connected by bus networks. It is a site in need of regeneration. The decline in industrial activities and freight services in SRSP has resulted in a significant reduction in economic activity on site. There are some industrial functions which continue to thrive (e.g. cement industry) alongside the commercial port functions (ferry and cruise services). Other industries have closed (e.g. the gasworks) leaving brownfield sites in need of decontamination and regeneration. Closure of these activities has resulted in job loss and economic deprivation amongst those communities remaining on the site. Alongside the need to regenerate the port, the Stockholm region is suffering from an undersupply of housing, particularly affordable housing (City of Stockholm, 2015).

SRSP benefits from good access to water and green space, to the east is the Baltic and to the north and west is the Royal National City Park. However, both land-based and aquatic environments have been degraded by industrial (contaminated land and disused infrastructure) and commercial port (emissions and waste-water produced by vessels) functions (Ports of Stockholm, 2018). Finally, SRSP (in common with the rest of Stockholm) has been suffering increasingly with problems of flash-flooding, probably resulting from climate change (Communication, 2016).

The municipality aims to regenerate the site over a 20-year period (Figure 3.3), integrating a liveable city district with industrial and commercial port functions. It also aims to connect the city district with the rest of Stockholm, enabling walking and cycling (City of Stockholm, 2017). Once complete 12,000 new apartments, 35,000 work spaces and 135,000 m^2 of commercial space will be constructed on site (ibid). It is hoped that the redevelopment will produce 30,000 jobs in port-related operations, financial services, media, start-ups, and the relocation of cultural services (City of Stockholm, 2018). The remaining port operations will be modernised. The container port and oil facilities will be moved from Loudden to Norvikudden. Housing will be built in its place.

Stockholm Royal Seaport aims to be fossil fuel free by 2030. This will result in a reduction of 36,000 tons of carbon dioxide per annum compared to business-as-usual. This will be achieved through the development of energy efficient buildings[2] and smart grid[3]; the use and production of renewable energy onsite[4]; waste management[5]; traffic and mobility management[6] and resource-efficient production. The new development will integrate the ecocycles system with low carbon transport (e.g. biogas buses, cycle and pedestrian networks, electric car-share schemes).

The plan is to move progressively towards the fossil-fuel-free target over the 20-year period. For example, the energy provider (Fortum) operating ecocycles will gradually increase the renewable content of the system to replace the fossil fuels used (Communication, 2016). Currently, the system uses biofuels (37%), waste incineration (31%) and fossil fuels (32%) (Stockholm City Council, 2015). Fortum plans to increase the quantity of woodchip (biofuel) imported into Värtahamnen to reduce the use of fossil fuels in the future (City of Stockholm, 2018). This will be complemented by the solar technologies installed on all new buildings.

The port also aims to be climate-adapted to rising temperatures, sea and groundwater levels, as well as increased precipitation (Ports of Stockholm, 2018). This goal will be achieved by raising ground levels on site and using local storm-water management provided by blue and green infrastructure integrated into the development (ibid). A Green Space Index will be applied to all new development, which identifies the optimal planting regimes, for regulating ecosystem services for storm-water management, biodiversity and recreational purposes (ibid).

The City of Stockholm is investing 130 million euros in the project (Communication, 2016). It is responsible for land remediation and infrastructure (e.g. streets, public spaces, cycle paths, bridges and park). It is also responsible for public engagement in the regeneration process (Communication, 2016).

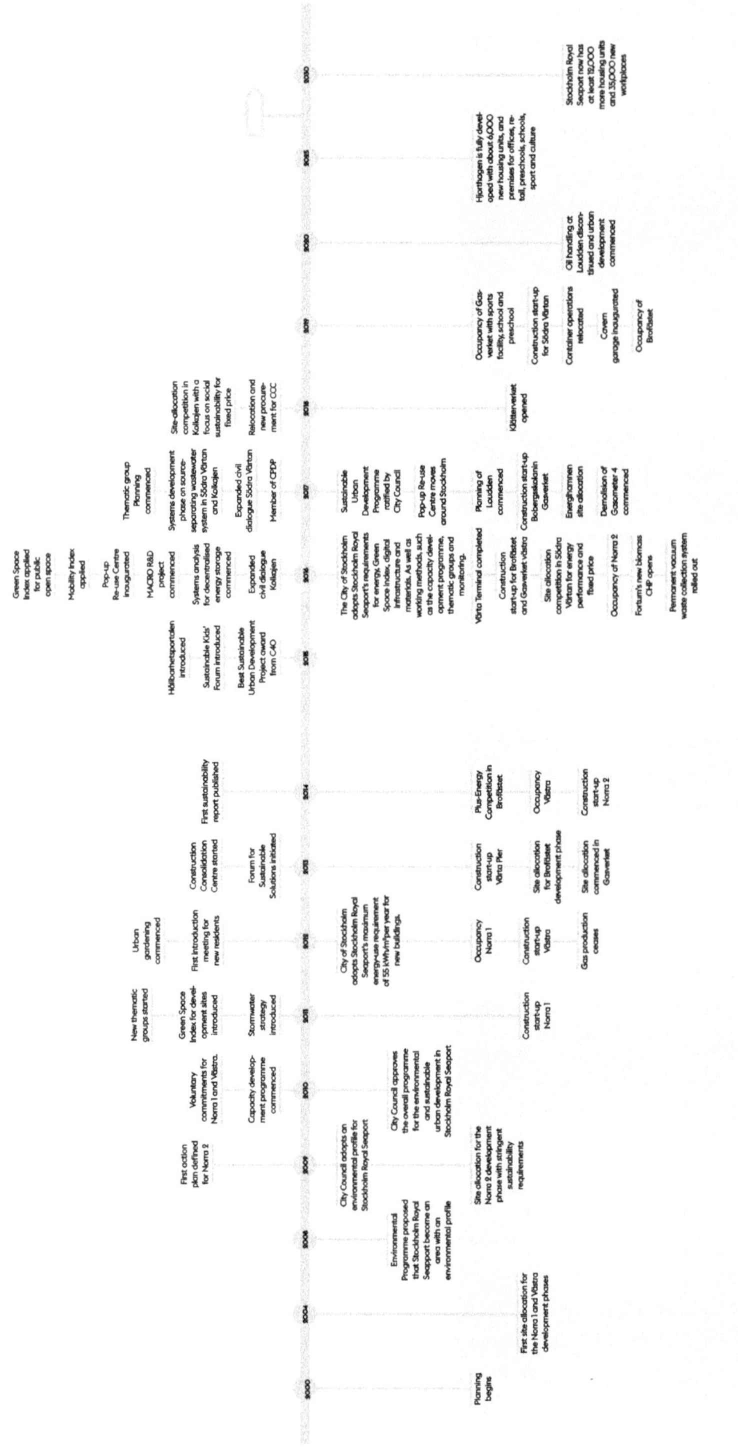

FIGURE 3.3 Regeneration time-line for SRSP.

Source: City of Stockholm (2018).

The public have been engaged in the planning process through a digital dialogue, an onsite open house, focus groups with selected audiences (e.g. business owners and young people) and the sustainable kids' forum (City of Stockholm, 2017). A collaborative planning approach engaging developers and service providers at an early stage in the development process (before the design competitions) has created a more integrated approach to provision and enabled the delivery of the stringent environmental targets set for the site (Communication, 2016).

Looping actions

Looping actions are being applied to the commercial port, industrial and living areas in SRSP. Ecocycles 2.0 operates across the port (Figure 3.4). It has been expanded to include additional waste generators (e.g. port functions, green infrastructure). Organic wastes produced on ships and from the maintenance of green spaces in the seaport and national park are used to feed the system. This reduces the amount of waste going to landfill and provides an energy alternative to fossil fuels. It creates compost which can substitute for fertilisers made with petrochemicals. It reduces the damage to the aquatic environment, the eutrophication of water-ways, caused by the release of waste-water into the harbour. A grey-water reuse system has been added to Ecocycles 2.0. The system stores storm-water in retention ponds or caverns, which limits flash-flooding in SRSP. Later, it is reused for watering vegetation in the port.

The whole lifecycle of the development has been considered as part of the ecocycles approach. Thus, the city has promoted circular construction and management processes in SRSP. One example is the minimisation of construction waste, by re-using materials on site. A second example is the treatment of contaminated soil for reuse on site. A third example is the recycling of garden and park waste within the port to produce compost and biofuel to be used on site. These actions close resource loops locally, thus reducing the need for transport.

Adaptive reuse of infrastructure using biological processes has also been adopted. Contaminated caverns on site (previously used for storing naphtha) were cleaned by filling them with water, then introducing archaea microbes to break down the naphtha (City of Stockholm, 2018). The waste products from the process (water, CO_2 and compost) were harmless. The caves are now used as a garage (for 1,200 cars).

The waste-water from vessels (black- and grey-water) creates the second biggest environmental challenge for the port (Ports of Stockholm, 2018). Waste-water can be offloaded at all quay berths in Stockholm. Since the 1990s the vessels operating routine scheduled services have offloaded their waste-water in port. Today, 98% of the waste-water generated by ferry passengers and 80% of the waste-water generated by international cruise ships are offloaded in port.

The waste-water is recycled to produce fertiliser and biogas (ibid). This reduces eutrophication caused by the release of nitrogen, phosphorous and potassium into the aquatic environment. The port authority also requires the solid

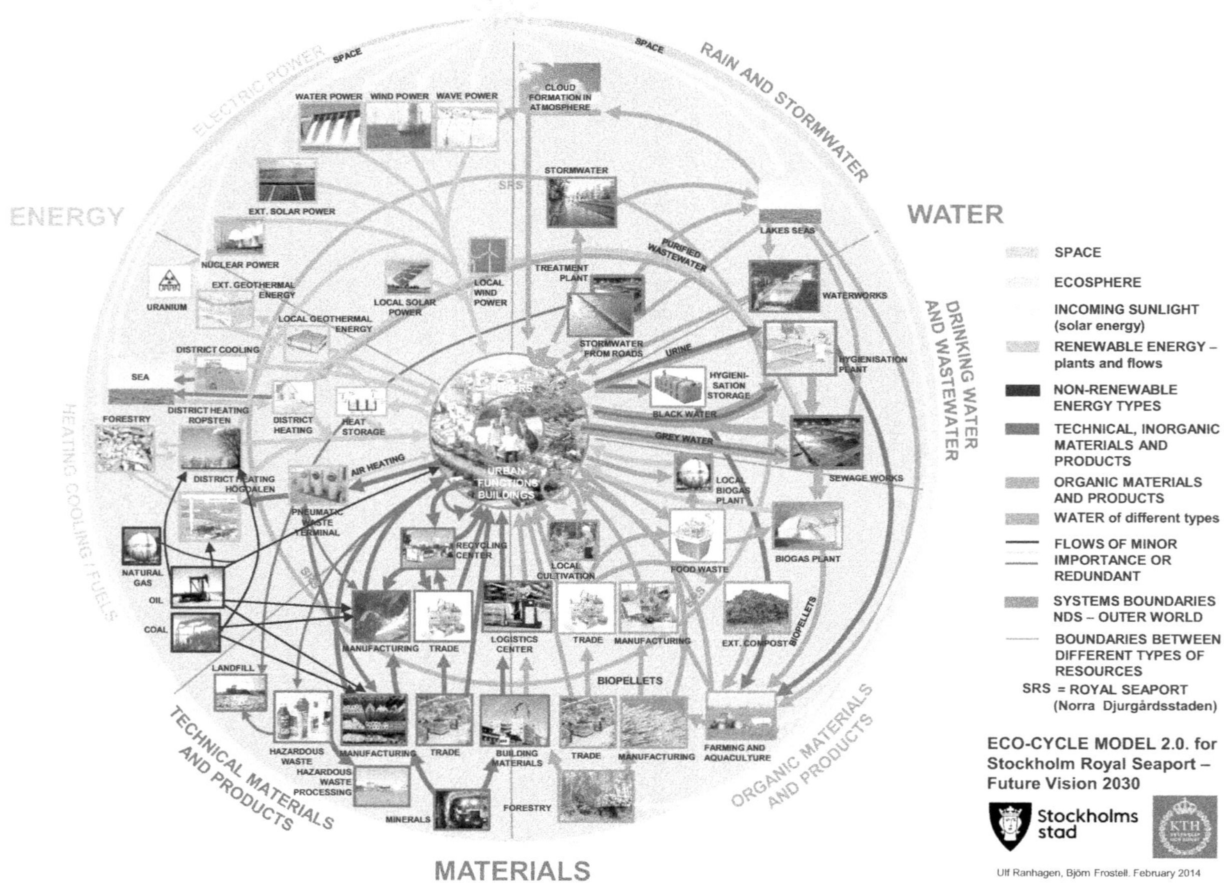

FIGURE 3.4 Ecocycles 2.0 in Stockholm Royal Seaport.

waste from ships to be separated (ibid). This enables the port authority to feed the waste generated into the closed-loop system operating in the area to produce electricity, heat, biofuel (for public buses), biogas for cooking and fertilisers.

The knowledge that is built in the port around circular construction and management processes is communicated to key actors: industry, government and the public. A lifecycle analysis tool[7] is being tested on site by developers (City of Stockholm, 2018). The tool helps to reduce resource consumption and waste in the construction process. The REFLOW model, based on ecocycles 2.0, visualises the port's hidden resource flows and demonstrates how these interact with local, regional and global flows of energy, water and materials. The tool is available online and is used to inform those living in or visiting the port area about resource flows and looping actions (ibid).

These looping actions create a cleaner, safer and healthier living environment for those working and residing in the port. They reduce the wastage of resources and avoid the environmental and economic costs of landfill, pollution and flooding. Looping actions will also generate jobs, for example, in waste collection and separation; energy generation and distribution; production and redistribution of compost and biofuel; soil and cavern decontamination; systems monitoring and training operatives. However, opportunities for industrial symbiosis, remanufacturing, reprocessing, recycling and reuse of waste materials/goods are currently under-developed in the port.

Ecological regeneration

The ecological regeneration of the site is also extremely important. Green and blue infrastructure will be integrated across the port to restore the land-based and aquatic ecosystems as well as reinforce the identity of the port (City of Stockholm, 2018). It will also create a healthy and attractive living environment. The introduction of blue-green infrastructure into the port will generate jobs associated with maintenance, conservation and recreational activities (Communication, 2016). It will offer a more attractive environment for tourists, which could generate further job opportunities in hotels, retail and catering (ibid).

Green rooves, courtyards and tree-planting have been integrated into new and existing development. It is estimated that green rooves and green courtyards will cover 14,000 m^2 and 28,000 m^2, respectively, when complete (City of Stockholm, 2017). The city plans that residents will have good access to parks and areas with high recreational and conservation values. The development plan requires that 100% of apartments have access to parks or natural environment within 200 m (ibid), which has mental and physical health benefits.

Green infrastructure will be connected across the site and will link with the national park and the waterfront. Green areas will be designed to be multifunctional to cope with future climate change, including storm-water management, to contribute to biodiversity and create good habitats (City of Stockholm, 2017). Gardening and urban agriculture on site will help return nutrients to the soil.

Vegetation will reduce noise and air pollution within the port, which has health benefits. The green infrastructure provides a dispersal network for oak-dependent species, pollinators and amphibians (ibid). This will increase biodiversity.

Blue infrastructure has also been integrated into the port. The aim is to reduce the impact of flash-flooding, which is a major problem (ibid). A mixture of permeable surfaces and retention ponds are being used. Storm-water runoff from streets and pavements is led to the surfaces of the planting beds by macadam mixed with bio char (ibid). Water is also led to lawns in the urban park, which serve as retention areas (ibid). The bio char is made from Stockholmer's garden waste and is ideal as a soil conditioner, and for capturing and binding CO_2 from the air (ibid).

Storm-water retention ponds prevent flash flooding, but they can also be used for watering vegetation and reducing pollution in adjacent waterways. Water bodies are being protected and extended in the port to strengthen and develop their recreational and conservation values. For example, in Kolkajen, a new island at the mouth of Husarviken creates a water-arena for residents and visitors (City of Stockholm, 2018). Thus, blue infrastructure protects grey infrastructure and the properties of those living in the port. It also provides opportunities for recreational activities which promote the health and well-being of those living and working in the area.

The Green Space Index (GSI) was developed and tested in Kolkajen and Södra Värtan. It is a tool for calculating eco-efficient space,[8] which rewards a range of ecosystem services. This has enabled developers to test the environmental benefits and economic feasibility of delivering different blue-green solutions for their projects in the port (Communication, 2016). It is used by city planners to ensure that adequate blue-green infrastructure is provided in new developments across the port. It has subsequently been integrated into the planning process across the rest of the city (Communication, 2016).

Adaptive actions

The adaptive capacity of the infrastructural systems and community has also been developed in SRSP. The physical planning and urban regeneration of SRSP is characterised by long-term robustness and flexibility (City of Stockholm, 2017). To make this possible, the area's zoning plans are flexible enough to accommodate a range of functions and future changes. Public buildings are designed to be multifunctional, to ensure optimal use (ibid). Public spaces are designed for different functions throughout the year. For example, a square could be used as a skating rink in the winter, a farmers' market in spring and an entertainments venue during the summer.

Public spaces are able to accommodate temporary events and activities during the construction period and when the area is completed. Some pop-up activities have already emerged, including a local market and a pop-up reuse centre (Communication, 2016). The pop-up reuse centre enables residents to recycle, repair or swap household items close to their home. It is now travelling around the rest of Stockholm, being used to test and build demand for such a facility (ibid).

Urban living environments also need to be able to adapt to changing demographic and environmental trends. Thus, the port offers a range of services for all ages (City of Stockholm, 2018). It also provides accommodation across a range of sizes (ibid). Thus, households of different sizes, at different life-stages, should be able to live in the port. However, the reality is the apartments are expensive. Thus far the development has largely excluded lower-income groups (ibid). The buildings and public spaces in the port have also been climate-adapted. This has been achieved though the elevation of buildings and public spaces, and the integration of blue and green infrastructure (ibid).

The adaptive capacity of those living in the port is being developed through their engagement in fora, workshops and resident groups. The public are engaged in decision-making processes (City of Stockholm, 2018 and Communication, 2016). Public understanding of problems and solutions develops through this engagement and builds support (ibid). It also creates networks through which the community can self-organise and learn. In addition, online apps have been developed to engage a wider audience in decision-making processes and provide learning platforms (ibid).

Stockholm Royal Seaport begins to provide us with a basic understanding of the dynamic relationships likely to emerge from a circular development pathway (Williams, 2019c). There are positive synergies between regenerating and looping actions in the seaport. For example, waste-water recycling (looping action) removes pollutants causing eutrophication in the local water-ways (regenerative action). Microbial remediation of naphtha (looping action) in the caves enables local storm-water storage (regenerative action). Land recycling and soil remediation (looping action) increases potential for green infrastructure to flourish locally (regenerative action). Storm-water captured by blue infrastructure (regenerative action) is reused for watering trees on site (looping action).

There are also positive synergies between regenerative and adaptive actions in the seaport. The provision of green and blue infrastructure (regenerative action) helps the environment to adapt to climate change (adaptive action), particularly by reducing flash flooding. Regenerated blue and green spaces (regenerative action) are used for pop-up activities (adaptive action), for example, sports and cultural festivals. Finally, there are also positive synergies between adaptive and looping actions. For example, the adaptive reuse of caverns (adaptive) is enabled by biological remediation (looping action).

SRSP also demonstrates some of the synergies between circular actions and other urban strategies. For example, localising activities helps to support looping actions by reducing costs of transportation. Construction, garden waste and contaminated soil are reused and recycled locally in SRSP. There is also a positive relationship between local activities and adaptation. Local engagement in collaborative planning, community fora and residents' associations has helped to build adaptive capacity within the local community (ibid).

However, conflicts have also emerged. For example, optimisation and substitution both reduce the potential for operating waste-to-energy systems (i.e. Ecocycles 2.0). By reducing waste generated within SRSP or increasing renewable energy production, the feasibility of operating the closed-loop system is reduced (ibid). Equally, the existence of the district heating system in Stockholm also reduces the impetus to substitute with renewable energy (Communication, 2016). By understanding the dynamics between these actions, we can begin to identify the combination of actions which are more likely to be successfully implemented.

Summary

Stockholm does not have a circular strategy. Nevertheless, it has applied circular thinking to urban development for several decades. Thus, the development regime has already transformed. Stockholm demonstrates two distinct pathways for circular development. The first pathway is a strategic, city-regional approach based on the *ecocycles* waste-to-energy system. The second pathway uses planned eco-districts to demonstrate and test the application of the circular actions in new build developments (e.g. Hammarby and SRSP). Both cases highlight challenges to implementation, which have emerged over a 20-year period. SRSP also demonstrates the dynamics between circular actions and with other urban strategies. Circular actions appear to broadly synergise with each other. However, there are conflicts between circular actions and other urban strategies, especially optimisation and substitution.

Notes

1 Funding streams included Local Investment Programme 1998, Swedish delegation for Sustainable Cities 2008 and Climate Investment Programme 2012.
2 New build energy requirements 55 kWh m^2/year for residential buildings.
3 A pilot smart grid system is being installed under a new model of collaboration between the private sector, academia and local government. This joint venture was additionally sponsored by the Swedish Energy Agency and the Swedish Governmental Agency for Innovation Systems.
4 Developers must install solar PV to cover 10–20% of building electricity need.
5 Zero waste to landfill target and the automated waste management system reduces energy use by 75–80%.
6 The mode split should include 70% of work-related trips by public transport.
7 The tool was developed by the Swedish Environmental Research Institute.
8 Space that makes a positive contribution to its own ecosystems and local climate, and to the social values associated with greenery and/or water.

4
CIRCULAR LONDON

Circular economy in the United Kingdom

At a national level the UK does not have a circular economy strategy. However, it does embed the principles of circular economy in the National Industrial Strategy. The aim of adopting this approach has been to increase resource efficiencies in the supply chain, production and disposal processes, creating economic savings for industries and business. Thus, the national focus is on circular economy, the modification of businesses models, production systems and restructuring of industrial sectors. The UK economy has grown over the past 20 years. However, the pace of growth has been relatively slow. The circular economy is expected to produce new business models, diversify the industrial base and generate jobs. It is estimated that the circular economy could be worth as much as £9–29bn a year (Eunomia Research Consulting, 2016) and create 10,000–175,000 jobs across skills levels by 2030 (Voulvoulis, 2015).

A laissez-faire approach to delivering circular economy has been taken by the UK government, which produces a rather fragmented picture. Some companies like Unilever, Kingfisher and M&S have integrated circular concepts into their business models. The British Standards Institution has created a voluntary framework for implementing the principles of the circular economy in organizations. The Environmental Audit Committee and Local Government Association have called for the public sector to lead by example and include circular requirements into public procurement policy.

Different approaches have emerged across the UK as a result of key policy areas being divided between different bodies, including Defra, BEIS, the Treasury, local authorities and devolved administrations. The focus of the circular transformation process is not spatial but sectoral in nature. Key sectors of interest are construction materials, plastics, food and textiles. In contrast to the rest of the

UK, the Scottish government has a comprehensive strategy for transitioning to a circular economy (Scottish Government, 2016). Zero Waste Scotland is responsible for delivering much of this. There is a Scottish Materials Brokerage Service for recycled materials and a Scottish Institute of Remanufacture.

At a national level there is no clear understanding of the role cities should play in the circular transformation of industrial processes. This may be in part due to a lack of a government department with responsibility for cities. Nevertheless, there are some British cities adopting circular economy strategies, the most prominent being London, Glasgow and Peterborough. The role for these municipalities in delivering their circular economy strategies is as a procurer of circular products and services or enabler in knowledge creation.

Circular London

London mirrors the policy taken nationally. Thus, the focus is on delivering a circular economy, by encouraging the creation of circular businesses and closing resource loops for specified industrial sectors. The motivations for this are economic (i.e. economic growth, diversification and efficiency) and environmental (reducing waste and emissions). It has been forecast that by 2036, the circular economy could provide London with net benefits of at least £7bn every year and 12,000 new jobs in re-use, remanufacturing and materials innovation (LWARB, 2017).

London has a waste problem. In 2016, local authorities collected 3,700,000 tonnes of waste. Despite attempts to reduce waste through reuse and repair, the total amount of waste generated in London has only slightly reduced over the last decade. More is being incinerated than ever before, and recycling rates have now dropped back down to 2010 levels. Yet London's population is predicted to reach 10.8 million by 2041. If Londoners continue to produce the same amount of waste, local authorities will have to collect an extra 1 million tonnes of rubbish each year. Experts have warned that this growth will be unsustainable and put an increasing strain on waste infrastructure, land and resources.

It has been estimated that moving to a circular economy in the capital would reduce material waste by 60% by 2041 (ibid). It is also expected to reduce greenhouse gas emissions from landfill and indirect energy consumption (transportation, production, disposal of products and materials). This could help London to achieve the target of moving to zero carbon by 2050 and zero waste by 2026 (ibid). The London's Circular Economy Route Map was produced by LWARB in 2017.[1] The strategy encourages the reuse, recycling and energy recovery from commercial waste, through the transformation of business models, supply chains, production and disposal processes. It currently focuses on five waste streams – construction, food, textiles, plastics and electronic waste – due to their volume and value.

The role of the GLA in this transformation process is in procuring circular products and services and enabling knowledge creation and sharing. The GLA

Group spends £11 billion per year on procurement activities (Greater London Authority, 2017a). It has recently updated its Responsible Procurement Policy (ibid), reflecting the need to procure circular economy services. It aims to incentivise the development of appropriate supply chains, expertise and business models. The Environment Strategy highlights the need for the GLA to show leadership in implementing circular practices, with the intention to influence other procurement bodies.

The Mayor also plans to support London's existing and future businesses to develop the skills, knowledge, experience and expertise to be competitive in a low carbon, circular economy. A clean-tech hub will be established in West London, providing workspace, collaborative opportunities and business support (LWARB, 2017). Mechanisms for financing start-ups and for knowledge sharing are being developed. Capacity building for a circular economy is mainly driven by the GLA, LWARB, WRAP and Ellen Macarthur Foundation (EMF) in London. The Circular Economy Club also provides a forum for discussion between businesses and exchange of good practice. However, funding for enabling actions to be taken appears to be limited. The focus here is on commercial waste and facilitating business to change its supply and production models on a voluntary basis.

There are two key waste streams which link circular economy to circular development in London. These are the construction and food waste streams. Tackling both waste streams has implications for land-use and infrastructure and thus for circular development.

Circular construction (looping and adaptive actions)

London's office space will increase by 5,000,000 m^2 by 2030, while over 40,000 units of housing will need to be built annually (over the next 10 years). Infrastructure development in the city is also a high priority, with CrossRail and the Thames Tideway Tunnel currently being built and extensions to the Northern and Metropolitan underground lines planned (LWARB, 2017). Construction, excavation and demolition waste constitute 48% of all waste in London (Greater London Authority, 2011). Tackling the construction waste stream could generate economic growth of between £3bn and £5bn annually by 2036 and 600 jobs by 2030 (Mitchell, 2015). The circular strategy prioritises the adaptive reuse of infrastructure (retain structures, refit or refurbish) over the recycling of construction materials (ibid).

The focus of the strategy is on encouraging the organic development of innovative business solutions in the sector. For example, the accelerator programme (run by LWARB) encourages the creation of small start-up companies and enables rapid scaling-up of innovations. The first cohort of start-ups focussed on modularisation, design for disassembly and reuse, technology to prolong asset life, alternative materials and material reuse, turning products into services and the creation of data platforms. The initial start-ups have designed an app for exchanging concrete (Sustainability Cloud); mushroom-based building insulation, which consumes waste as it grows (Biohm); a material for use in building and interior architecture produced from agricultural and food waste (Organic Refuse Biocompound); modular housing, with component tagging and tracking and an online marketplace where components can be traded as part of a circular economy. However, a more strategic, city-regional approach to monitoring, localising and looping construction waste flows appears to be absent.

The Draft London Plan (Greater London Authority, 2017b) has identified 38 opportunity areas (including the Queen Elizabeth Olympic Park and Old Oak Common) as places that will see unprecedented levels of regeneration and development. These projects generate demand for new materials, while the demolition of existing buildings creates large volumes of waste that are typically "down-cycled" to lower grade products. Requirements for supporting the circular economy have been integrated into the London Plan, alongside green infrastructure and climate adapted urban form. This links circular economy directly with circular development. Demonstration projects will be supported by the GLA and local authorities. Indeed, circular principles have already been embedded into the redevelopment of Queen Elizabeth Olympic Park.

The Circular strategy aims to build capacity within the construction industry to deliver circular development. Training workshops have been funded by LWARB. A study to identify underutilised buildings in London and encourage greater reuse of vacant spaces is underway. A network of facilities and office managers has been created to integrate circular principles into building operations and enable knowledge sharing. Local authorities are also encouraged to use temporary spaces for circular activities.

Smaller scale demonstration projects are beginning to emerge in London, which test building designs for adaptive reuse or recycling. One example is the PLACE/Ladywell in Lewisham. It is London's first pop-up village, designed for disassembly and re-assembly. It provides temporary accommodation for people on the housing waiting list and space for co-working. It has been constructed on a vacant site. The buildings comprise prefabricated, modular, highly energy efficient pods, designed to be dismantled and relocated within the borough when the whole site is redeveloped. This approach provides one potential solution to the housing crisis in London, whilst reducing construction waste.

Circular food (looping, ecologically regenerative and adaptive actions)

Over 8 million tonnes of food is consumed in London per year by the city's 8.6 million residents (GLA, 2015a), around 1 million daily commuters (GLA, 2015b) and almost 17 million annual overseas tourists (ibid). Approximately 20% ends up as waste. More London boroughs (70%) are collecting food and green waste than in the past (ibid). However, London urgently needs to introduce or extend food waste recycling in its high-density housing stock. With tightening budgets, local authorities are often guided more by financial than by environmental concerns. Separate collections reduce food waste, but they are more expensive.

London also urgently needs new treatment facilities for organic waste (GLA, 2015b). The GLA aims to increase anaerobic digestion of food waste, to produce methane, which can be used in energy generation, heating and transport. It is estimated that the anaerobic digestion of food waste could save 175,000 tonnes of CO_2e emissions and £120,000,000 in disposal costs per annum by reducing waste going to landfill (ibid). Currently, less than 50% of London's food waste is processed in the capital (ibid).

There are only two plants which convert food waste within the M25 – Riverside Anaerobic Digester (Croydon) and London Sustainable Industries Park (Dagenham). Both use food waste to produce energy and Riverside also produces biogas. There are also four large-scale composting facilities.[2] As a growing city, London will require more facilities to process 1 million extra tonnes of food and green waste. The Mayor is encouraging the development of further waste treatment facilities; however, financing these facilities is a problem. The GLA could potentially offer some of its landholdings to enable, or directly provide, waste infrastructure, such as food waste processing plants.

The GLA has set a target to reduce food waste by 50% by 2030 (GLA, 2018b). This will lead to a significant reduction in the core materials associated with food packaging,[3] which collectively make up around 75% of municipal waste. The circular strategy advocates prevention, reuse (i.e. redistribution to those in need – foodbanks, animal feed), recycling (i.e. composting and anaerobic digestion) and energy recovery (i.e. combustion). It also advocates local food production as a way of closing resource loops locally (i.e. using food waste as compost). The latest estimates for London's net benefits from circular food economy opportunities are that they could add £2–4 billion annually to GDP by 2036 (LWARB, 2015).

London has a food strategy and various schemes to reduce food waste (e.g. Social Supermarkets, Food Save, Trifocal) and encourage food growing projects (e.g. Capital Growth, Incredible Edible). The London Food Strategy (GLA, 2018b) advocates the allocation of space by local authorities for urban farming in the capital. The GLA has highlighted the importance of including food growing spaces in new housing developments and as a temporary use on vacant or underutilised sites in the New London Plan. It also encourages local authorities to protect existing food growing spaces including allotments and promote urban greening in their local development plans.

The GLA provides funding towards the Capital Growth programme and supports other food growing networks to help promote the health, economic, environmental and community benefits of food growing. The funding will help maintain the network of over 2,500 food growing projects. The GLA recognises the contribution that food growing plays encouraging social enterprise and job creation in the food sector (e.g. Sustain's Roots to Work programme).

The GLA also works with organisations to develop bids to the Good Growth Fund, which is committing £70,000,000 investment by 2021 to support regeneration in London. The fund supports investment in the emerging nature-friendly urban farming sector, helping London to become a pioneer in urban agriculture and circular economy. This is further supported through public procurement contracts with local urban farmers. Urban farming is helping to ecologically regenerate London. It also increases the communities' ability to adapt to food shortages by increasing sufficiency. Like circular construction, it is central to a circular economy, but also affects the circular development pathway in London.

Circular development in London

The London Plan is the most appropriate vehicle for operationalising circular development. It sets out the three circular actions (looping, ecological regeneration and adaptation) separately, but does not indicate the synergies between them. It allocates land for "low value" circular activities (e.g. for urban farming, storage and logistics; for secondary materials and waste management and green space). It also supports the provision of sustainable infrastructure (e.g. heat networks, renewable energy, grey-water recycling and rainwater harvesting systems), green-infrastructure (e.g. green roofs, pocket parks) and adaptable infrastructure within strategic developments (e.g. the Olympic site). The London plan requires circular construction principles are adopted for strategic sites (GLA, 2017b). It encourages boroughs to support opportunities to use vacant buildings and land for flexible and temporary uses. The reuse of vacant properties is also encouraged through a tax on empty homes and dwelling management orders. The links between waste, water and energy systems are made in the London Environment Strategy 2018, which could help to encourage the creation of integrated closed-loop systems (as in Stockholm).

However, there are challenges to adopting a circular development pathway in London. The GLA is largely reliant on the UK government to create a regulatory framework and provide the funding required to enable the transformation process. However, the UK government has largely dismantled regulatory frameworks (e.g. zero carbon homes code) and funding mechanisms (e.g. green deal) which could have helped to support circular development. The GLA has limited powers and resources at its disposal. It has limited funds to leverage a circular transformation. But it does have land. It also has responsibilities for transport, waste and spatial planning. The GLA's powers include regulation, procurement, enabling and provision. In its regulatory capacity, it can set policy targets and produce policies to guide development, the economy and the environment. However, funds are extremely limited. Options for financing sustainable infrastructure (including integrated water, energy and waste systems) are being investigated. One suggestion is to use public pension funds to support the investment. This was a strategy adopted in Sweden to successfully fund the development of 1 million energy efficient homes during the 1970s (Williams, 2012).

Public procurement will be the main lever for driving circular transformation of businesses in the capital. There will also be some limited funds for capital investment and capacity building for SME's provided by LWARB. Limited financial resources mean the GLA is heavily reliant on other agencies to deliver the circular transformation. The GLA has an enabling role in this process, through knowledge sharing and capacity building. It also has a role in coordinating actions. The number and diversity of actors involved in the delivery of services, infrastructure and the management of resources in the capital create problems with coordination. This is particularly an issue when attempting to create integrated, looping systems, which by their nature cut across infrastructure (e.g. energy and water), services (e.g. waste, water, energy, transport) and industrial sectors. These actors operate within different legal frameworks, across different administrative areas with different goals. This creates problems in coordinating actions and aligning goals.

To illustrate how the circular development manifests in London, two contrasting case studies have been chosen: Queen Elizabeth Olympic Park (East London) and Brixton (London Borough of Lambeth). The former provides an example of a top-down approach to a circular development. It is a large-scale intervention, which uses a combination of regulation (i.e. planning and contractual agreements) and significant public funding to stimulate the circular transformation of a new urban district. The focus is on innovation in design, technology and industrial processes. In contrast, the Brixton case relies heavily on the community, social enterprises and Lambeth council to deliver circular development. The process is localised, and comprises low-tech, small-scale interventions in the existing community, through a series of grass-root actions. The focus here is on changes in social practices, alternative systems of provision and localisation of resource flows.

Circular Queen Elizabeth Olympic Park

The Queen Elizabeth Olympic Park (QEOP) is a major regeneration project in Stratford, East London. It covers 226 hectares, encompassing land in the London Boroughs of Newham, Waltham Forest, Tower Hamlets and Hackney (Figure 4.1). Once complete, the project will generate 40,000 jobs and provide 24,000 new homes. In 2005, the site was identified for the London 2012 Olympic Bid. The bid was successful and the Olympic Development Agency (ODA) was formed to implement the preparations for the Olympic Games. At the heart of the bid was the need to deliver a sustainable games and in the longer term to create a sustainable district in an area suffering from multiple deprivation and environmental degradation. Strategies to deal with demolition and construction waste; temporary structures post-games and attracting long-term, sustainable activities to the site were needed. Thus, a circular approach to development was required.

Originally, the site was fragmented, with a lack of amenities and limited employment opportunities for the local population. The wards in the Lower Lea

Valley were generally within the 10% most deprived in England. These areas had high unemployment, poor health and high crime rates. There was also a severe shortage of affordable housing locally, yet some social housing estates, travellers' community and the largest housing cooperative in Europe were removed to make way for the Olympic development. Environmentally the site suffered a range of problems. Land uses were generally low-value, interspersed with vacant and derelict sites, creating the appearance of neglect.

Historical usage of the site included oil refineries, chemical works, cold storage facilities, landfills and backfilled reservoirs, car compounds and warehouse/distribution centres. These uses created significant problems with soil and groundwater contamination. The waterways in the area had also deteriorated, having become silted up and overgrown. Moreover, the combined sewer system had insufficient capacity to handle storm-water discharge during peak time; consequently, two overflow points in the area discharged untreated effluent into the River Lea waterway system. The site was fragmented by numerous highways, railway lines and waterways, making accessibility within the area difficult. However, there were a diversity of habitats and species on the site.

During the early phases of development, the focus was on the construction and hospitality industries. Systemic changes were largely achieved through large-scale public funding and application of new environmental standards in contractual agreements, actor engagement in the design process and value added to their business. Post-2012, the London Legacy Development Corporation (LLDC), a mayoral development corporation, was formed to support the ongoing development of the site. It had spatial planning and compulsory purchase powers. It also had a limited budget, relying on public–private partnerships to deliver major projects.

The regeneration project incorporated all three circular actions. Looping actions were integrated into the development process (e.g. recycling construction materials, soil recycling) and operation of the park (e.g. recycling of food/ park waste, recycling black-water and reusing grey-water). The development was designed to be adaptive. For example, the site provided meanwhile spaces for pop-up activities. Temporary structures were built for the games, which were disassembled and relocated post-2012. The ecological regeneration process initiated in the park pre-games (to improve the soil and waterways, create a range of habitats and increase biodiversity across the site) continued post-games. This has increased the area's ecological carrying capacity and developed the ecosystem services. QEOP's designation as a park continues to support this process.

Looping actions

To ensure the environmental sustainability of the Games, one of the ODA's goals was to reuse, re-purpose, or recycle 90% of the soil and material waste at the site. A cut-and-fill strategy was deployed to use excavated materials in foundations beneath several sport venues. This reduced the carbon footprint that vehicles

FIGURE 4.1 Queen Elizabeth Olympic Park.
Source: Artistic impression produced by Sally Williams.

would have generated by moving contaminated soil from the area. The Olympic Park became the UK's largest ever soil-washing operation (Atkins, 2012). Contaminated soil was treated using a variety of techniques including bioremediation, soil washing, and chemical and geotechnical stabilisation. Two "soil hospitals" were set up on site (to test, process and treat excavated contaminated soil for reuse). Five soil-washing plants were used to treat a range of contaminants. In the Olympic Park 80% of the soil was cleaned and reused for the enabling works (ibid). Beyond the Olympic Park, the soil hospital linked up with projects such as the M25 widening works and the neighbouring Westfield retail development in Stratford to maximise reuse of surplus materials, thus avoiding dumping in landfill.

The recycling of black-water was also trialled in the park. Thames Water established a 7-year R&D project (which ended in 2019) centred on the provision of grey (non-potable) water from the Old Ford Water Recycling Plant (OFWRP). The plant reclaimed waste effluent (black-water) from the main sewage outfall (the Northern Outfall), and processed it to grey-water standard. It then supplied the surrounding areas through a dedicated network to serve a range of current uses including toilet flushing, irrigation and water cooling. It has supplied on average 75 million litres a year of recycled, grey-water, which reduced the use of potable water by the same amount on site annually. The substitution of grey-water for common water usage activities helped to achieve significant reductions in potable water usage, and provided additional drought resilience for the parklands (LLDC, 2017). The project cost around £7,000,000 and was funded by the ODA and Thames Water.

Thames Water was contracted to operate the OFWRP and supply grey-water to customers to February 2019 (Thames Water, 2019). However, the scheme ended in 2019. There was uncertainty about potential demand for non-potable water in the QEOP. There were difficulties securing long-term funding for the waterworks, and the financial return for grey-water was significantly less than for potable water. Also the potential for cross-contamination between potable and non-potable systems created a barrier. Thus, Thames Water decided to focus on grey-water recycling and rainwater harvesting systems in new developments. These would be paid for by the developer and property owner, rather than Thames Water. In fact, according to the Environment Agency's flood attenuation advice (LLDC, 2017), rainwater harvesting was likely to be limited on site. Nevertheless, two developments have rainwater butts in their gardens (Sweetwater and East Wick) and a further development has adopted grey-water recycling (Chobham Manor Exemplar Homes).

The ODA also set an ambitious target to reuse or recycle 90% of the demolition and construction waste produced by the development, which it exceeded. Designing out waste by embedding goals in design briefs, procurement policies and contractual agreements was critical for delivering the reductions. Providing guidance to design teams to indicate how materials could be reused or recycled or to design for future disassembly was critical. Integrating pre-demolition audits with materials management planning was key to reducing demolition waste going to landfill. Waste reduction produced cost savings for the contractors. There was one centre for the management of all construction waste produced on site. This allowed waste to be aggregated and dealt with at scale, thus reducing the effects of transportation, reducing costs and increasing potential markets. Accurate forecasting to enable effective resource management was essential (identifying resource quantities and potential markets). Award schemes to incentivise waste reduction further were also implemented. Approximately 98.5% of construction and demolition waste was recycled, 425,000 tonnes of waste were diverted from landfill and 20,000 lorry movements were saved (LLDC, 2017). Also 22% of the aggregates used on site were from recycled and secondary sources. It is estimated over 20,000 tonnes of new materials were saved through this approach (ibid).

Adaptive actions

Circular thinking was woven into the design and building of the Olympic Park, its venues and facilities. The park provides an example of how, by designing for adaptability from the outset, buildings, products and materials can continue to provide high value return beyond their initial purpose. An Asset Disposal scheme was set up to help contractors reuse items and materials after the Olympic Games by selling them or gifting them to charities and good causes. For example, the nine modular cabins that formed the "High Street" in the Athletes' Village are now used as a community hub in Hackney Wick. The warm-up running track, which was laid without tarmac so that it could be easily removed, was donated to British Athletics. Colourful rubber matting used on bridges and walkways throughout the park was reused at a primary school in Northern Ireland, and lamp columns were donated to a local skate park. The International Broadcast Centre was designed with a flexible internal layout so it could be reused for many different activities post-games. The MacDonald's restaurant on site was a modular construction. This allowed it to be disassembled and reused on another site after 2012.

Meanwhile spaces for temporary pop-up activities were also integrated into the development, through the LLDC's strategy for interim uses. The land was parcelled for different types of temporary uses including green infrastructure, commercial uses, servicing the park and some grass-root activities. Commercial temporary uses included a range of events, such as seasonal fairs, theatre performances, circuses, sports competitions, or less visible commercial lettings, such as open storage or car parking. These activities were expected to help revitalise the park, before phased residential development began. Some of the pop-ups proved popular and have remained. For example, the View Tube built from recycled shipping containers is a temporary café. It also provided educational resources for community engagement in the redevelopment of the Olympic Site. The commercial success of the café, combined with the limited alternative uses for the site, means the lease has been extended indefinitely. The LLDC produced guidance for grass-root pop-ups (LLDC, 2014), which identified and demarcated sites on the west side of the QEOP, alongside the Lee Navigation Canal, for projects. In the surrounding neighbourhoods, the LLDC has encouraged a range of temporary uses, from community allotments to pop-up shops (Leyton) and from roof gardens (Stratford) to community centres made from recycled construction materials (Hub 67, Hackney Wick). Alongside the more directly managed interim use programme, these uses have been commissioned and supported through calls for proposals and small-scale funding, such as the "Emerging East Commission" and a programme known as "launch pads". Thus, an adaptable urban fabric has been created.

Ecological regeneration

The ecological regeneration of the area has also significantly benefitted the local ecosystem (Figure 4.2). The QEOP is the largest urban park to have been created in Europe over the last 150 years. The blue-green infrastructure it provides has helped to tackle urban heating and flooding; increased local biodiversity; helped to re-establish natural cycles (e.g. nutrients, phosphates, nitrogen) and enabled carbon sequestration. It has also produced health benefits for those living and working on the site. Green space of at least two hectares can be found no more than 300 metres (5-minute walk) from park residents. This benefits both their mental and physical health. Green corridors have been constructed throughout

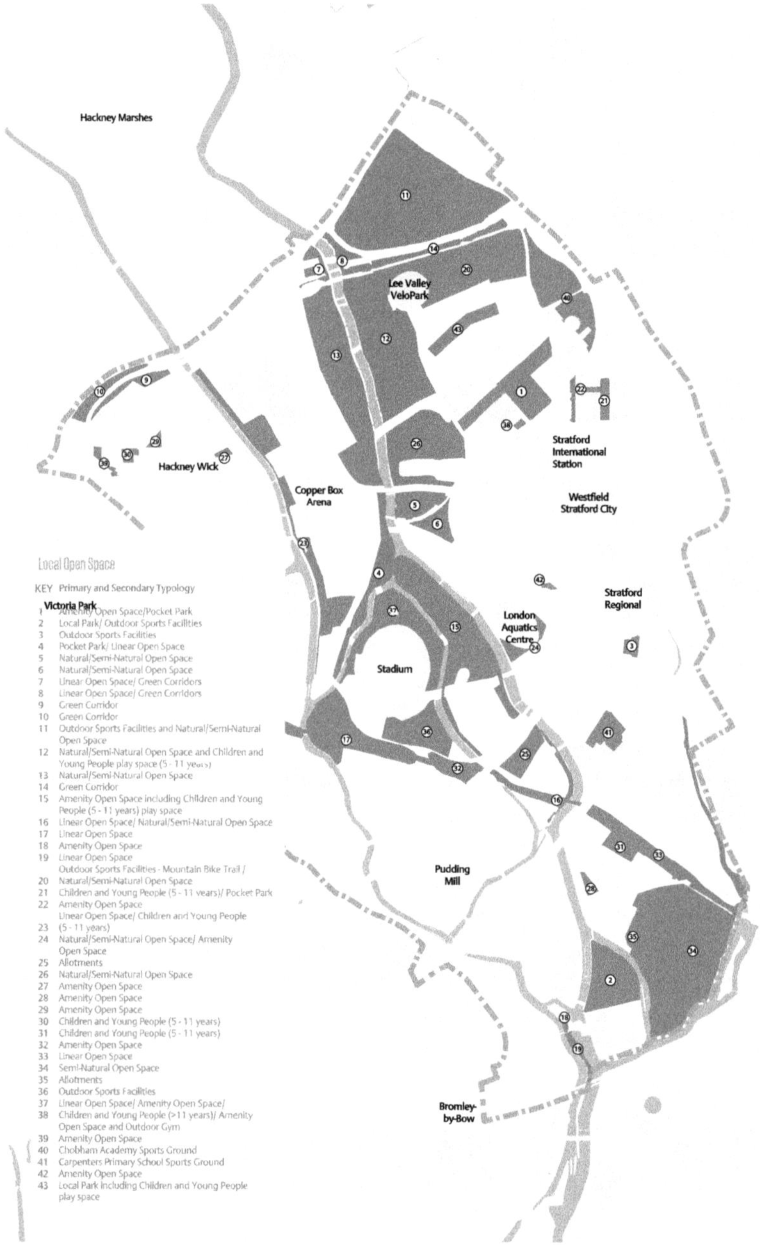

FIGURE 4.2 Green infrastructure.
Source: London Legacy Development Corporation (2020).

the park. One green corridor is more than 40 km in length, following the River Lee connecting the Green Belt (outer London) to the Thames (inner London). Diverse, natural species have been planted across the park, to increase biodiversity. The green infrastructure index (also used in Stockholm Royal Seaport) is used by developers on site (e.g. Chobham Manor, Sweetwater and East Wick developments) to determine the planting strategies likely to deliver the greatest ecological benefits in new developments.

The QEOP also includes 6.5 km of improved waterways. The River Lea flows through the centre of the park and is an attractive ecological asset to be preserved and enhanced. Several streams diverge around Stadium Island. There are opportunities for further ecological enhancement of these waterways through additional embankment stabilisation. In the North Park and Canal Park areas, sustainable urban drainage components (i.e. rainwater gardens, swales, retention ponds, wetlands) are being fully integrated into the public realm. This does not apply for most areas of the South Park due to the additional flood risk management requirements.

There is a problem with ground water contamination on site. A new bioremediation technique was applied for the first time in the UK to treat groundwater contaminated with ammonia beneath the Olympic Stadium. *Archaea – naturally occurring* micro-organisms that thrive in extreme conditions and that biologically degrade ammonia – were inserted into boreholes. As Archaea removed the ammonia, several other reagents, including oxygen-released compounds, were injected into other groundwater areas to remove contaminants. Thus, ground water was decontaminated (a process also used in Stockholm Royal Seaport).

The QEOP has acted as a test-bed for a range of processes, design solutions and technologies which are fundamental to a circular development pathway The new processes in place, for reducing construction/demolition waste and for soil recycling, have proved extremely successful. These processes are now being employed on large construction sites across London. In contrast, the recycling of black-water to produce grey-water although extremely successful in terms of reducing the consumption of potable water has not proven financially feasible. Sadly, this will not be adopted more permanently or widely across London. Yet in an area of drought, this looping action should be integral to the system. The problem seems largely to result from the valuation of potable and non-potable water. This example highlights the need for public funding to support such ventures which are in the wider public interest, until infrastructural and valuation systems catch-up.

There have been significant economic and ecological gains from adopting a circular development pathway on this site. However, there have been losses socially. The ejection of the original communities from the site and lack of opportunities provided by the new development for the deprived communities is a significant failing. It should be addressed through policy, regulation (e.g. the planning system), public engagement in decision-making, education and funding. Of course this failure is not integral to the circular development pathway, but it is endemic in regeneration projects. In a circular urban system existing groups should be retained to preserve local social capital and a systems adaptive capacity. Thus, it is essential to have a diversity of employment opportunities, housing tenures and types which cater to the needs of existing and future inhabitants.

Circular actions can produce a diversity of jobs which should help to support a diverse economy and variety of people (again helping to develop the systems adaptive capacity), but this will require skills training (and not necessarily from the elite institutions which are currently planned for the site). Flexible, low cost, low carbon housing which provides warm, affordable accommodation for the low-income groups is also required. Potentially the provision of pop-villages (similar to Lady Well) on site, as a meanwhile use, could also help with the wider housing crisis affecting London.

Circular Brixton

Brixton is an area within the inner London Borough of Lambeth. It has a very diverse population. Today, it is a borough divided by considerable wealth and economic deprivation. During the 1980s–1990s Brixton had several periods of social unrest. Over the past 20 years, it has been going through a process of rapid regeneration, but disparities still exist. In 2007, Lambeth Climate Action Group suggested Brixton should become a Transition Town. There was an increasing public interest in combating climate change and creating a new development trajectory post-peak oil. The group aimed to tackle climate change, increase local self-sufficiency and address local social issues.

Brixton became a Transition Town in 2010. Brixton Transition Town became a community interest company (social enterprise). This was the first stage in a process which led to the adoption of a circular development pathway in Brixton (Figure 4.3).

A local currency – the Brixton Pound – was introduced in 2009 to localise resource flows (including monetary flows), reduce emissions from transport and reinforce the local economy. To be successful it was critical that a substantial amount of the currency circulated and could be used to pay for goods and services locally. The currency could be spent at over 250 businesses in the area. Business rates could also be paid using the currency which incentivised take-up. By 2012 Lambeth Council also paid part of their employees' wages using the local currency. Usage further expanded in 2013 when an electronic version was established, and customers could pay by text. This combination of actions reinforced the circulation of the local currency.

It was hoped that the local currency would also encourage local circular actions. Some of the businesses in the scheme engaged in circular actions. For example, the Remakery sold upcycled products, Brixton Pound Café reused food waste, Brixton Energy generated renewable energy and Loughborough Junction Farm grew food and used compost produced from local organic waste. Some businesses were based in buildings which had been adaptively reused (Remakery),

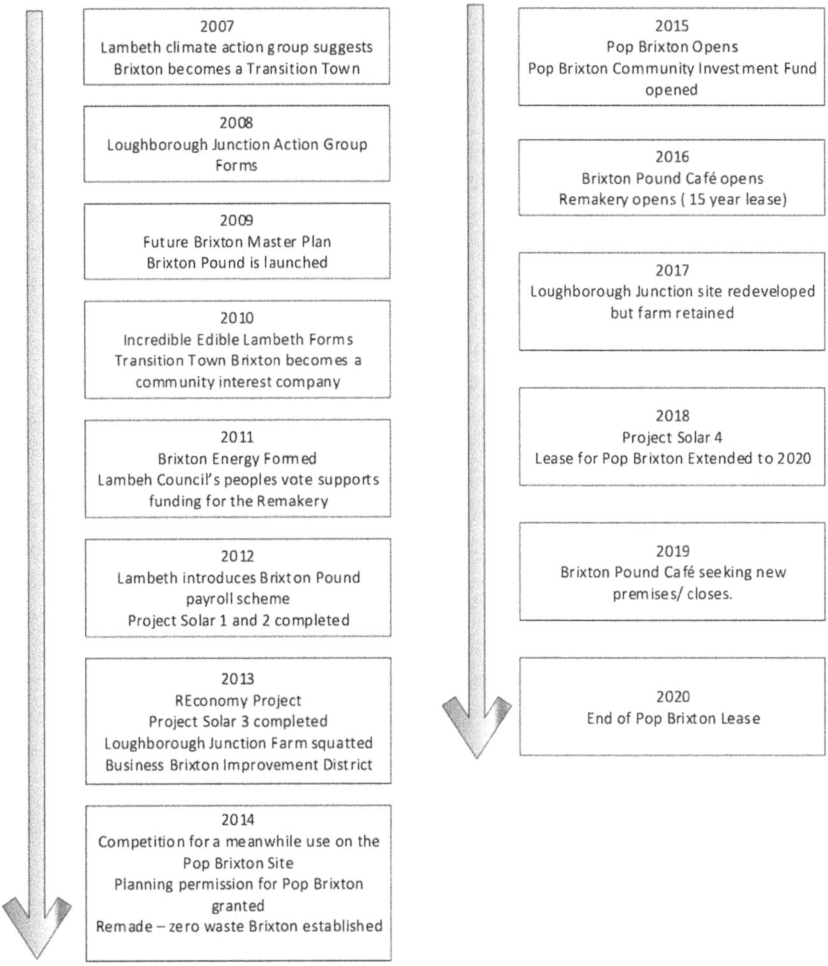

FIGURE 4.3 Brixton time-line.
Source: Author's own.

were built from recycled materials (Pop Brixton) or were on sites which had been reused. Thus, purchases made using the Brixton pound supported these circular activities. Nevertheless, some of these activities failed because they were not commercially viable (e.g. Brixton Pound Café).

Alongside the local currency, the key focus for Transition Town Brixton was energy. Brixton Energy was formed in 2011. It aimed to decarbonise the local energy supply and increase the efficiency of local buildings. It worked in partnership with REPOWER London. Local projects for generating solar energy and energetic refurbishment appeared across the borough. The focus was on retrofitting renewable and efficiency technologies in social housing (enabled by Lambeth Council) to address green-house gas reductions and fuel poverty. Four solar projects have been completed to date. Residents generating renewable energy were paid in the local currency, to encourage spending within the local area. Energy could also be bought using Brixton Pounds. Thus, the local currency supported the generation of renewable energy and the local circulation of energy resources.

Transition Town Brixton also wanted to develop a local food system. Local food production in Lambeth proved popular. It was supported by Lambeth Incredible Edible, which was founded in 2010. Loughborough Junction Farm emerged in 2013, after a community action group squatted a contaminated brownfield site (Figure 4.4). Local food production had both ecological and social benefits. It offered valuable ecosystem services in a densely developed borough (ecological regeneration). It also provided the opportunity for local residents to develop valuable horticultural skills and build community networks (building adaptive capacity).

Initially, produce was grown in bags to avoid contamination from the soil. Organic waste from the local café and growers was used as compost on site. The food produced was sold in Brixton market and used in the community café. Thus, the food-loop was closed. This was one of many local growing projects in Brixton. In 2017, the site was redeveloped for commercial and residential uses. However, the farm was retained at the centre of the development and was subsequently given a 20-year lease. This project provides an example where looping, adaptation and ecologically regenerative actions combine.

Projects for encouraging the recycling and reuse of goods, materials and infrastructure also began to emerge in Brixton. The Remakery was one such initiative

FIGURE 4.4 Loughborough Junction Farm.
Source: Artistic impression produced by Sally Williams.

instigated by Transition Town Brixton as "Remade Brixton". The Remakery opened in 2016. It was based in a reclaimed underground car park. It offered training for the local community, enabling the reuse and recycling of a range of goods. It was awarded capital funding from Lambeth Council in 2011 and has a temporary lease for 15 years. The idea behind the Remakery was to develop local skills in repair and refurbishment, to reduce waste and create employment opportunities. Co-working spaces were offered to small businesses, start-ups and artists. Members could access the materials, skills workshops and tools needed to repair and refurbish products. Any profit subsidised membership and social programmes for disadvantaged groups in the community. The project enabled the looping of materials and increased the adaptive capacity of the local community.

A further grass-roots development – Pop Brixton – also embodies circular actions. The project was completed in 2014. The buildings were fabricated from recycled shipping containers, designed for easy disassembly and reassembly. It offered cultural and employment opportunities for the local community. It also established a community investment fund offering workshops, training and even work experience for local people. The café used locally grown food, helping to close the food-loop locally. The project enabled the looping of infrastructure and land, alongside building the adaptive capacity of the local community (Figure 4.5). Pop Brixton has been a success and received extra funding from the GLA. However, its lease was extended only until 2020.

FIGURE 4.5 Pop Brixton.

Source: Artistic impression produced by Sally Williams.

Brixton offers an interesting case study. It takes a grass-root, temporary experimental approach to circular development. The combination of a local currency, a mixture of activities and the close proximity of producers and users of "waste" resources are essential to its success. The projects have had limited support from the GLA. However, the local authority has offered support through allocation of land/property on temporary leases; through the adoption of local currency for paying council workers and business taxes and also through some limited funds allocated by the people of Lambeth. Brixton offers an example of how an existing community might retrofit circular actions, with limited funds, relying less on technology and more on *people-power.* These small, low-tech interventions have proved successful in terms of their social and ecological objectives. However, the long-term economic viability of some projects is precarious (e.g. Brixton Pound Café, Remakery, Loughborough Junction Farm). This is due to heavy reliance on volunteers and the temporary planning permissions. In such a climate, the commercially viable activities run by Repower and Pop Brixton seem more likely to persist.

Summary

London does have a circular economy strategy. It is focussed on the creation of circular businesses and industrial sectors in the capital. Nevertheless, through the London plan all three circular actions are addressed (albeit separately). Thus, London does offer examples of circular development. It demonstrates two distinct pathways. The first pathway uses a planned eco-district to demonstrate and test the application of the three circular actions in a new build development (QEOP). The second pathway adopts a grass-root, temporary experimental approach to delivering circular actions in an existing neighbourhood (Brixton). Both cases highlight some of the levers (e.g. planning, public funding and procurement) for circular development. They also highlight that economic challenges create the greatest barrier to circular development in London.

Notes

1 A statutory board established by the GLA Act 2007 and chaired by the London Mayor.
2 Ecopark Compost Centre (North London), Rainham Compost Centre (East London) West London Composting and Sutton/ Mitcham composting centre (South London).
3 Paper, card, plastic, glass, tins and cans.

5

CIRCULAR AMSTERDAM

Circular economy in the Netherlands

The Dutch government introduced a Circular Economy programme in the Netherlands in 2016, with the aim to transition by 2050 (The Dutch Ministry of Infrastructure and the Environment and the Ministry of Economic Affairs, 2016). The programme focussed on the decoupling of growth from material use and on a system in which the sustainable extraction of raw materials and the preservation of natural capital were guaranteed. This is essential if the Netherlands is to address its biocapacity deficit (−4.2 gha/capita[1]). The recognition of the deficit by the national government is important, because it means that the development of a circular economy is clearly linked to domestic ecological regeneration.

The Circular Economy programme mainly focusses on changing business models and production systems to reduce material consumption. However, water, energy and infrastructure (as a potential source of construction materials) are also highlighted as resources which need to be managed. Thus, the circular economy strategy touches on all the resources central to the circular development, with the exception of land. In addition, the Netherlands Environmental Assessment Agency indicated that spatial planning solutions could contribute to the transition to a circular economy. Thus a link between land-use, urban form, infrastructure and material flows has been acknowledged.

The Dutch have several motivations for adopting a circular economy. The first is resource security. The Netherlands imports 68% of its raw materials from abroad (ibid). To become more resource secure it aims to reduce imports, by reducing consumption and resource wastage and encouraging the use of renewable or ubiquitous resources. The circular economy programme aims to ensure that raw materials in existing supply chains are utilised in high-value activities. This can lead to a decrease in the demand for raw materials. Where new raw materials are needed, fossil-based, critical and non-sustainably produced raw materials are replaced by sustainably produced, renewable and generally available raw materials.[2] This preserves natural capital and future-proofs the economy by making the Netherlands less dependent on the import of finite sources. The Dutch aim to develop new production systems and promote new forms of consumption, giving impetus to reduction, replacement and utilisation. This results in greater sufficiency, building adaptive capacity and reinforcing national resilience to resource scarcities.

The second motivation is the contribution circular economy can make to national growth. The Netherlands has a good starting position to capitalise on circular economy. It has good infrastructure, transport connections and strong relevant industrial sectors (including the chemical industry, the agro-food sector, high-tech systems and materials, logistics and recycling), all of which build capacity for the circular transformation. The Netherlands also leads the way when it comes to the bio-based economy and the utilisation of nature-based solutions that reduce the use of raw materials. Dutch design is setting trends internationally and the government intends to take a leading role for circular design as well. TNO states that an extra turnover of €7.3 billion can be generated annually by the circular economy, producing 54,000 jobs in the Netherlands (Bastein et al., 2013). This will not only affect the industrial base in the Netherlands but also influence the provision of infrastructure and services in Dutch cities.

The third driver is mitigating climate change. A more responsible use of raw materials fits with the Dutch climate policy. The annual emissions released in the Netherlands are close to 200 megatons of CO_2 equivalent. An improvement in efficiency in raw material and material value chains could cut this by approximately 17 megatons of CO_2 a year (which is 9% of total Dutch emissions) and thus make a significant contribution to achieving the climate objectives (Blok et al., 2017). At the same time, nature-based solutions (e.g. green infrastructure) for climate mitigation and adaptation help to reduce the demand for primary raw materials (e.g. grey infrastructure for drainage systems) and promote the transition to circularity (e.g. grey-water recycling). The Dutch realise that securing natural capital will contribute to solutions in both domains. However, the realisation of climate and energy goals will increase the demand for some raw materials for renewable energy technologies (generation, storage, and transport). A circular economy is also important to meeting this demand. The Dutch government recognises that circular economy could be good for public health and the environment. Designing products in such a way that they can be fully reused and recycled or can be safely released into our environment as ecologic raw materials will have social benefits.

The Dutch definition of circular economy is far broader than that adopted in the UK. The government takes a holistic view of circular economy (looping), recognising the linkages with nature-based solutions and natural capital (ecological regeneration); adaptive and resilient systems, specifically in relation to resource scarcity and climate change (adaptation). It asserts the environmental, economic and social benefits of adopting a circular approach to development. It also recognises the dynamic links between materials, water (especially waste-water), energy (waste-to-energy), infrastructure (as a source of materials) and land-use. The recognition of these synergistic relationships is embedded in policies and strategies at a local level.

Circular Amsterdam

Amsterdam, together with its surrounding municipalities, acts as an economic driver for the region and country. Its dynamic, service dominated economy includes both major international firms and small start-ups. The city-region is well connected to the rest of Europe. It has a large international airport and port. Amsterdam has grown considerably since 1890. Present population trends show consistent growth forecast for both Amsterdam and the metropolitan area over the next 25 years. The expectation is that that the city will grow from 834,713 in 2016 to just over 1,000,000 in 2040. This projected population increase of approximately 23% to 2040 comes upon sustained population growth over the past 15 years.

The population increase is putting pressure on the municipality to provide living accommodation and associated services within the metropolitan area. The municipality has created residential islands – reclaimed land – in the bay. City planners estimate that there is currently enough space to meet the growing housing demand by transforming unused or underused spaces such as former industrial sites. This regeneration process might provide an opportunity to transform the city's infrastructure and services, helping it to achieve the sustainability goals set out in the strategy. However, increased densities and urban expansion can negatively impact on the urban ecosystem and ecosystem services. This could potentially exacerbate the problems of flooding already experienced in the city. Sea-level rise will create further problems in future years. Amsterdam is also heavily dependent on gas-powered district heating systems. The city wishes to transform the energy system, moving towards 100% renewable power, to increase energy security and help to mitigate climate change.

In 2015, Amsterdam adopted a Sustainability Agenda (Municipal Council Amsterdam, 2015). This sought to tackle the decarbonisation of the energy supply, air quality, climate mitigation and adaptation. The Sustainability Agenda planned for the management of a range of resources (land, water, materials and energy) and infrastructure. Central to the agenda was transitioning to a circular economy. The motivation for adopting the approach was to establish Amsterdam as a pioneer in delivering circular economy and thereby gain economic advantage.

Amsterdam has many entrepreneurial and innovative citizens, start-ups, research institutions and companies that are working on the circular economy. The first bio-based and circular clusters of mutually supportive businesses already exist in the port. The municipality has encouraged innovation and circular activities as part of its active contribution the national commitment to the Netherlands becoming a "circular hotspot". Amsterdam is collaborating with regional municipalities, the Amsterdam Economic Board and numerous other partners to create a circular economy at the regional scale, so that acceleration and upscaling can be achieved. The city's long-term ambition is to create a circular economy with new methods of production, distribution and consumption.

The sustainability agenda made linkages between circular economy, ecological regeneration and adaptation. It took an integrated view, which underpinned the delivery of a circular development. The Sustainability Agenda proposed Amsterdam would be a testing ground for circular district development and circular economic activity. Thus, it distinguished between circular economy and circular development. It also stated that Amsterdam would be the first Dutch municipality to develop a large transformation area, using circular development principles.

Early in 2016, the spatial and economic development plan for Amsterdam Metropolitan Region was published (Figure 5.1), which embedded the principle of transitioning to a circular economy in the development process (Amsterdam Metropolitan Region, 2016). The municipality launched two ambitious circular programmes: *Amsterdam Circular: Learning by Doing* and the complementary *Circular Innovation Programme* in 2016.

The *Learning by Doing Programme* produced 20 circular projects for the municipality, including procurement and land development (City of Amsterdam, Circle Economy and Copper8, 2017). It utilised the municipalities planning and procurement powers in combination with municipally owned land to enable innovation. It aimed to use the experiments to prove that circular development could be profitable, thus encouraging wider adoption by the development regime.

The *Circular Innovation Program* encouraged the municipality to work with businesses and knowledge institutes to deliver the circular economy. Overall, 30 innovative projects were developed, including circular start-ups (ibid). The municipalities' role here was largely one of enabling. It partially funded projects, but mainly helped to facilitate a transition to the circular economy. For example, the municipality supported capacity building within industries through the development of knowledge networks.

The *Learning by Doing Programme* produced the *City Circle Scan*. The scan monitored resource flows in the city-region. It demonstrated that two value chains were very important: the building and construction sector and the biomass and food sector. It was estimated that the material savings made by adopting looping actions for both waste-streams could add up to nearly 900,000 tonnes per year (Bastein et al., 2016). This was a significant amount compared to the existing annual import to the region of 3.9 million tonnes. It was estimated that this saving could reduce CO_2 emissions by 500,000 tonnes annually (ibid). Taking these actions would affect the way in which the city-region was planned and developed.

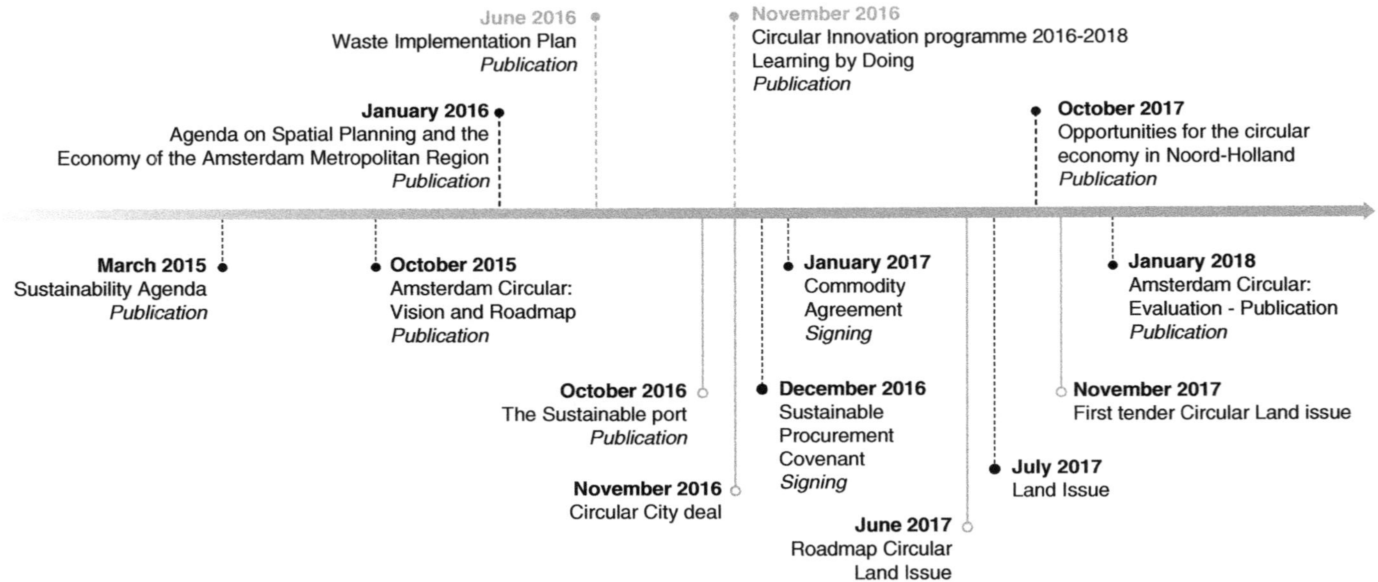

FIGURE 5.1 Time-line for circular policies in Amsterdam.
Source: City of Amsterdam, Circle Economy and Copper8 (2017).

Circular construction (looping and adaptive actions)

The first circular development pathway focusses on construction waste. The Netherlands already recycles 98% of building materials annually. Amsterdam is a leader in this activity. Several companies based in Amsterdam are specialised in the management of demolition and construction waste. A key role is played by the AEB, a world leader in the sustainable conversion of waste into energy, precious metals and reusable raw materials. AEB recycles about 61,400 tons of materials annually (especially ferrous and non-ferrous metals), reducing CO_2 emissions by approximately 172,500 tonnes/year (AEB, 2015).

The Circle Scan platform provided information about flows of construction and demolition waste in the city-region. It showed there was potential to increase the reuse and recycling of both. It demonstrated that through the implementation of material reuse strategies there was potential to create a value of €85 million per year and 700 jobs in circular construction (Bastein et al., 2016). However, the economic feasibility of recycling or reusing construction materials and components relies on localising resource loops (Figure 5.2). Potential producers and consumers would need to be located within the city-region, in order for exchange to be cost-effective. Due to the time-lag between demolition and construction projects, storage facilities would also be needed to facilitate the process. Both would have land-use implications and could be supported by spatial planning.

In addition to the Circle Scan, an online platform – PUMA – identifying the presence of high-value resources in the built environment was developed. Project PUMA provided a geological map of Amsterdam showing the presence and availability of high value metals (iron, copper and aluminium), in the built environment across the city-region. It also explored the possibilities for extracting the metals from the urban mine. It provided valuable information for those demolishing infrastructure, enabling them to determine the potential value of the materials salvaged. It also provided information for those wishing to reuse the salvaged materials in Amsterdam. Data platforms such as PUMA are essential for the effective exchange of reused or recycled resources and the creation of local supply chains in city-regions.

Amsterdam launched its first roadmap for circular buildings in 2017 (Metabolic and SGS Search, 2017). It challenged the private sector to develop circular buildings and circular city districts. Four circular strategies were proposed. The first was high value reuse and recycling. This involved the repurposing of buildings, components and upcycling of materials for new building products. The second strategy was smart design, which produced flexible, adaptable and recyclable buildings (e.g. Hubbell in Amsterdam builds adaptable, modular spaces). Both strategies recognised the link between looping and adaptive actions and could be reinforced through the planning process. The third strategy enabled the exchange of resources between producers and users through the provision of a physical resource bank and an online digital marketplace. The fourth strategy

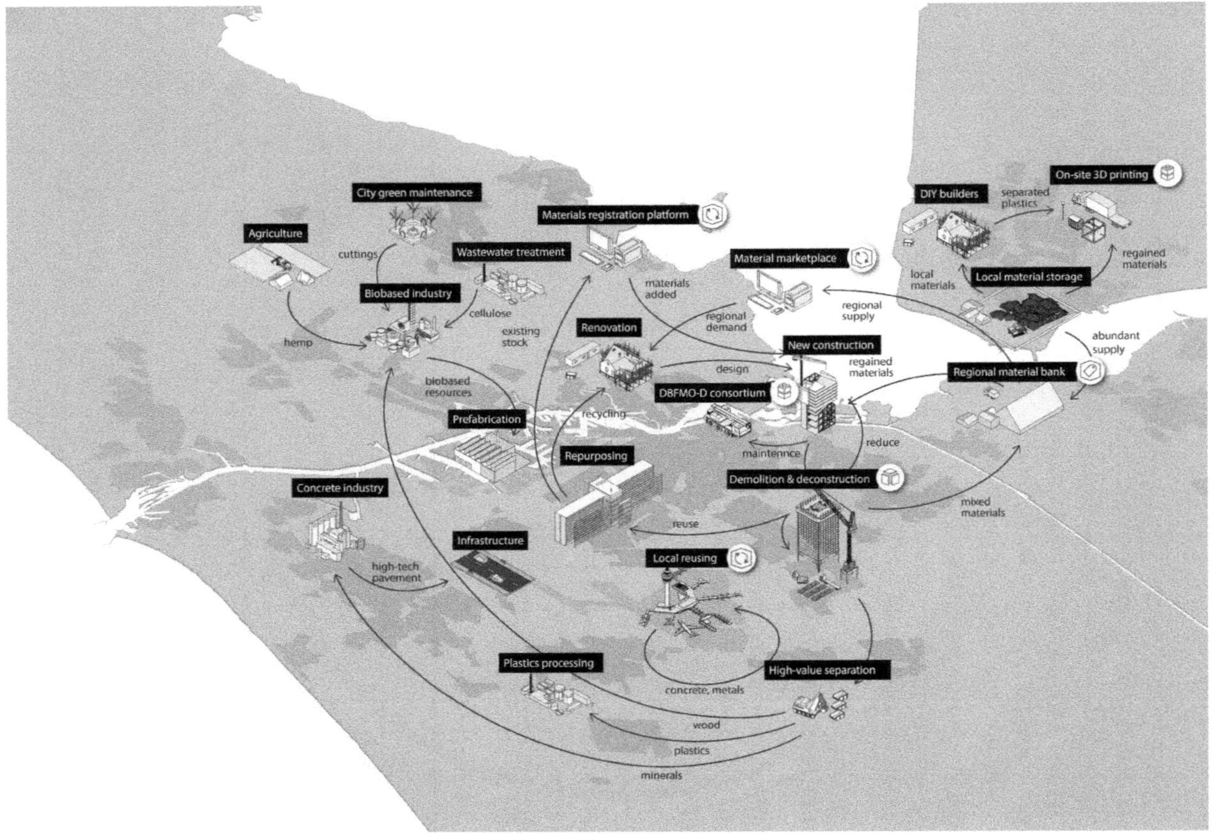

FIGURE 5.2 Vision of a circular construction chain for Amsterdam region.
Source: Bastein et al. (2016).

improved the separation of waste streams, enabling components from dismantled buildings to be reused or recycled more easily.

Circle scan analysis

The Circle scan analysis was completed to identify strategies which could encourage looping of construction materials, components and buildings in the city-region. It determined the economic value and employment generated, the material and CO_2 savings for each strategy (Table 5.1). This informed which projects would be supported by the municipality. Smart design was forecast to produce the greatest carbon-dioxide savings. High value reuse was predicted to produce the greatest material savings. Dismantling infrastructure, separating materials, creating an online market place and providing resource banks were likely to generate the most value.

Circular tendering and land issue

The projects completed through the *Learning by Doing Programme* demonstrated that market actors were willing to build in a circular way, as long as demand was demonstrated (City of Amsterdam, Circle Economy and Copper8, 2017). Demand could be generated by the city through planning and procurement policies. Planning policies requiring the high-value reuse of buildings, components or materials or the construction of smart buildings created demand amongst developers. Procurement policies also generated demand for locally sourced, recycled or reused materials and components. For example, the municipality

TABLE 5.1 Impact of four circular construction, biomass and food waste strategies

	Construction				*Organic waste*			
	High value reuse/ recycling	*Smart design*	*marketplace and resource bank*	*dismantling & separation*	*cascading organic flows*	*bio refinery hub*	*waste separation and logistics*	*recovering nutrients*
Value creation (euro millions)	23	12	25	25	30	30	50	30
Jobs	200	100	200	200	450	450	200	150
Material savings (Ktonnes)	500	0	0	0	75	75	300	500
CO_2 savings (Ktonnes)	75	300	25	100	300	300	100	100

Source: Bastein et al. (2016).

required that reused baked bricks were used to construct 100% of the public realm works in the city. In this way the city helps to stimulate the circular transformation process.

Circular land issue was also pivotal in this process. It is an instrument used for tenders for urban transformation, infrastructure renovation and demolition. Circular tendering applied circular criteria to the release of public land or buildings for development across five categories: materials, energy, water, ecosystems and resilience. These criteria recognised the important linkages between looping, ecological regeneration and adaptation. There were three motivations for developer engagement in the process: first, to gain access to public land for development (often at a reduced cost); second, to develop their expertise in the arena of circular construction and third, to demonstrate their sustainability credentials.

Since 2017, circular criteria have been successfully applied to four development tenders for public land in Amsterdam: Buiksloterham, Centrumeiland, Zuidas and Sloterdijk. The first circular tender was completed in Zuidas in 2017 for a large project (250 homes and offices). This included the use of material passports and dry connection practices to enable future reuse and recycling of built structures. Secondary (recycled) materials were also used in the construction for insulation and partition walls. Important lessons have been learnt from these circular projects, which can be used to inform the planning and development processes (City of Amsterdam, Circle Economy and Copper8, 2017).

The first lesson is that variation in local characteristics produces unique projects. Projects are most successful when both generic and area-specific goals are formulated early in the development process. By providing area-specific goals in the development plan, the process can be streamlined. Second, it is important to identify the scale (regional, neighbourhood and building level) at which circular goals should be implemented. Businesses need this focus in order to formulate their own goals. A clear designation of responsibilities for delivering goals, at different scales, is essential for facilitating the process.

Third, the complexity of the tendering process increases for circular development, because teams need to make choices between circular goals. This may take time to resolve and creates tensions in large-scale circular procurements. Thus, a circular project requires a different approach from the municipality as well as from businesses. The planning teams must allow more time for realising circular ambitions. Planners will also need to become experts in circular development, in order to advise developers effectively.

Fourth, prescribing functional tender criteria (such as adaptability and modularity) helps to prevent a future decrease in the value of infrastructure. It is important to build knowledge about suitable construction practices and materials, which enable circular demolition and disassembly, amongst construction and demolition firms. Demolition and disassembly plans in tenders for construction projects are needed to ensure that the lifecycle of the infrastructure is considered from the start of a project. A proactive approach from demolition companies in recovering value from residual materials in demolition projects is essential, if circular construction is to be successful.

Amsterdam has now adopted a city-wide policy for circular tendering. It aims to contribute to the development of a national standard for circular building. Already new networks for knowledge transfer to enable the development of circular construction practices have emerged. For example, a concrete network advocates the use of granulated, recycled concrete in new infrastructure. Living labs (e.g. FabCity, AMS and AUAS LivingLab) help to demonstrate circular construction methods, thus providing vehicles for learning.

Circular Innovation Programme

The *Circular Innovation Programme* has also spawned a variety of projects in the port of Amsterdam, which actively promote the recycling/reuse of construction waste (City of Amsterdam, Circle Economy and Copper8, 2017). One example is the Circl building. Its wall insulation contains clothing fibres, while the roof is made from 16,000 pairs of old jeans. On the roof, photovoltaic panels and a garden for water recovery have been set up. All the bricks and the tiles come from recycled material. This is a demonstration project which uses building information modelling technologies to assess material flows. It supports the cataloguing of all materials used in the structure. This information can also be used at demolition phase to enable the reuse of materials and components.

Developers share their knowledge from these innovative projects. For example, the developer OVG has actively shared knowledge on the circular transformation project – Edge Olympic Amsterdam.[3] The municipality has organised training for building contractors focussed on circular procurement and construction processes. Thus, knowledge and supply chains for circular construction have begun to develop in the city. The municipality also chairs the Circular Economy Task Force of Eurocities and is engaged in the C40 cities network. This enables Amsterdam to share its findings with other cities internationally. Through the use of urban experiments and knowledge exchange, Amsterdam seeks to drive the circular transformation of the construction industry locally, nationally and internationally. It also demonstrates what circular development could look like in practice.

Circular organic flows (looping, ecologically regenerative, adaptive actions)

The second circular development pathway adopted in Amsterdam focusses on organic waste. The circle scan suggests that the implementation of biomass and food reuse strategies in Amsterdam Metropolitan region has the potential to create a value of €150 million per year (Bastein et al., 2016). It could also create 1,200 additional jobs in the agriculture and food processing industry.

Circle scan analysis

The scan identified four strategies for enabling circular flows of organic waste in the city-region (ibid). All four impact on the way in which the city develops (Figure 5.3). The first aimed to improve waste separation and smart reverse logistics in order to valorise residual streams. The second strategy aimed to create cascading organic waste flows. This is to ensure the residual flows retain their highest value. In the case of food waste, this would mean selling surplus food in restaurants or reusing cooking oil as vehicle fuel.

The third strategy involves the development of bio-refineries in the city region, to enable organic materials to be recycled or recovered locally and at scale. This would produce biogas, compost, medicines, nutrients and chemicals. If processed locally the cost of transporting the organic material is minimised, which increases the economic viability of reprocessing and recovering energy. The fourth strategy involves nutrient recovery from residual food, for reuse (by restaurants or foodbanks) or composting. This would capture 95% of the nutrients lost currently.

The scan showed that greatest material savings would be produced by better waste separation and nutrient recovery from residual food (Table 5.1). The most economic value would be generated by waste separation and reverse logistics. The greatest carbon savings and the most jobs would be created by cascading organic flows and the local provision of a bio-refinery. Adopting these looping strategies would increase the potential for Amsterdam to become more self-sufficient in terms of energy production and food, and thus increases its adaptive capacity.

The biomass and food chain in Amsterdam is often closed in a low-value manner. Currently, 30% of the organic waste produced in Amsterdam is incinerated to generate electricity and heat. This is partly due to restrictive regulations. For example, unconsumed food products must be treated as waste, which makes high-value application difficult. It is also because until recently organic material had not been separated from the residual waste stream. However, as part of the plan to encourage high-value looping, Amsterdam has overhauled its waste collection service and now provides separate containers for organic waste.

Bio-based industrial cluster

The *Circular Innovation Programme* has encouraged collaboration between businesses, knowledge institutions and public organisations to deliver new bio-composite products from waste biomass. The municipal authority's role in the innovation programme is mainly supporting research, promoting information exchange and providing some financial support for projects. It has also designated sites for the development of the bio-industries in the Port area.

Closing organic waste flows will need to be supported by development. Clustering actors who produce and consume organic waste within the city will enable this. Thus, a waste cluster has been established in the Port of Amsterdam, comprising the Amsterdam waste-to-energy company (AEB) and water company (Waternet). AEB has an ambition to convert organic household waste into more valuable materials. Thus, it has invested in a post-separation facility, which aims to separate 65% of domestic waste for efficient recycling by 2020 (Port of Amsterdam, 2018). AEB is also looking to tackle waste produced after incineration and conversion of organic waste to biogas or animal feed by 2035 (ibid).

Waternet manages the water supply in Amsterdam Metropolitan Region. It is strategically located next to AEB in the Port of Amsterdam for optimal synergies of waste and feedstock. It is adopting cascading and nutrient recovery strategies. Waternet and the Amstel, Gooi and Vecht Water Board have been recycling phosphate from sewage water since 2013 in a phosphate factory in Amsterdam-West. With the phosphate from the Amsterdam waste-water, 10,000 football pitches can be fertilised annually. Waternet has also found that alginic acid can be recovered from granular sludge and used in the pharmaceutical or food industry (Van der Hoek et al., 2016).

Waternet has developed processes to generate biogas (for cooking, electricity and heat), fuels and advanced chemicals from sewage. It has developed techniques for separating cellulose fibres (from toilet paper) to produce building materials, paper products and bioplastic (ibid). Waternet and AEB are also collaborating on a project called Power-to-Protein. This project extracts ammonia from sewage to create high value proteins, sufficient to provide all the residents of the Amsterdam with 35% of their primary protein requirement (ibid). AEB provides Waternet with surplus electricity for bacterial production of protein.

There is also a bio-refinery cluster in the port (ibid). It is one of the largest bio-refinery clusters in Europe. It produces over 25 million m^3 of biogas, 5 megawatts of electricity and heat and 5,000 tons of fertiliser from organic waste (ibid). There is a 20-hectare site designated for a new bio-refinery in the port and warehouse facility dedicated to bio-based companies (Prodock). This facility provides space and a platform for start-ups and investors. Two companies are already established in Prodock: PeelPioneers[4] and Chaincraft[5] (ibid).

The port has significant storage capacity for biofuel (e.g. the storage tanks of Oiltanking). It also has a biodiesel plant and a direct kerosene pipeline to Schiphol International Airport (ibid). AEB has a steam pipeline which links to several sites in Amsterdam (ibid). Thus, there is infrastructure which could potentially be used to store and distribute biofuel from the port. Several companies on site (NWB, Koole, CWT and Cargill) are specialised in bio-based liquid or dry bulk logistics (e.g. sugars syrup, ethanol, veg oils and biomass). Amsterdam area has a long-standing tradition in chemical innovation and R&D, which may also prove useful in establishing bio-based industries in the port hub.

Localised food loops

In Amsterdam, urban farmers and restaurants are also searching for ways to valorise their organic waste. The RE-ORGANISE project aims to create knowledge and business solutions around decentralised production of (semi-finished) products, materials, water and energy from organic waste. Instead of paying for waste removal, urban farmers and restaurants aim to create new business opportunities by separating and processing their organic waste into valuable products on site. These products are either re-used by the producers or sold to partners nearby. By valorising organic waste in a decentralised manner, these actors aim to produce high-quality products and reduce sourcing, transportation and waste management costs, while gaining independence. However, logistics, financing and unclear or restrictive regulations are still obstacles for scaling up.

There are various schemes to valorise food waste in Amsterdam. For example, the *Too good to go* app is a platform enabling restaurants to publicise left-over meals to potential consumers. *"Taste before you waste"* reuses 250 kg of food on a weekly basis, turning it into plant-based meals at their *Wasteless Wednesday Dinners* or given to the *Food Cycle Markets* in Amsterdam. *Instock* is a non-profit restaurant that creates meals from unsold products from supermarkets. It has also established a *Food Rescue Centre*, a *Food Waste School Program* for primary schools, and a cookbook designed to encourage the reuse of food waste throughout the community. Organic waste is also used to create compost for local urban farms. For example, the Secret Village (a small commercial development) in Amsterdam makes compost from sewage for local urban agriculture projects. Urban agriculture provides an excellent opportunity to close the food loop locally, regenerate the urban ecosystem and increase food sufficiency (adaptive capacity). There are many projects in Amsterdam: herb gardens, city farms, vertical and cooperative farming.

Similar practices also happen at scale. In 2011 the energy company Meerlanden set up the Green Energy Factory (Groene Energiefabriek) in Rijsenhout, just south of Schiphol (Amsterdam Airport). It used the organic waste (food, agricultural and garden waste) from nine municipalities and 4,000 companies in the region to produce green gas, CO_2, heat, compost, citrus fuel and water (Amsterdam Economic Board, 2018). A digester processes 53,000 metric tons of organic waste; 60% is used to make biogas (ibid). This powers more than 50% of Meerlanden's vehicles; the remainder is used by households and industry in the surrounding area. The CO_2 is captured from the biogas and delivered to various local horticulture companies, which use it as a growth enhancer. Recently, Meerlanden has also started applying a technique to extract oil from citrus peel, producing a fuel that is used instead of diesel in their own weed control equipment.

Meerlanden has an innovative tunnel composting system, which yields 2.5 million bags of compost annually (ibid). A large proportion of the heat produced during the process is captured, totalling around 10 million kWh. More than a quarter goes to a nearby greenhouse horticulture company, enabling it to make a saving of 320,000 m^3 in its gas consumption (ibid). The composting process also produces 4.5 million litres of condensation (ibid). This water is used for street cleaning and anti-icing brine on the roads. The Green Energy Factory demonstrates how organic waste loops can be closed within city-regions. The collaboration with companies in the region and municipalities is the basis of the Green Energy Factory's success. The company receives financial support from the municipalities in which it is active, and a grant (Stimulation of Sustainable Energy Production grant) from central government. There is also income from the sale of the green gas, CO_2, heat, compost and water, or at least savings are made when these resources are used internally by the company.

The designation of land in the city-region for bio-refineries, compost storage, urban farming, farmers' markets, bio-digestion, waste-water treatment and protein production as well as the connective infrastructure will greatly influence Amsterdam's ability to build a thriving bio-economy (Figure 5.3). Both will have a profound influence over the way in which the city develops. The strategic planning process can be used to support this. However, the designation of space in Amsterdam for circular actions is potentially contentious. It is a dense city, experiencing considerable pressure to provide housing. Outside the city competition for space is equally fierce. The demand for recreational areas is expected to rise by 30% to 2040, creating conflicts between tourism and agricultural uses.

There are also conflicts between commercial agriculture and the protection of ecosystem services. Large-scale farms that serve the global markets in the peatland meadow areas (which are often of great cultural and historical significance) do not combine well with the spatial and environmental goals of maintaining an open countryside, ensuring sustainable water management, preserving biodiversity and reducing CO_2 emissions. Thus, land-use will need to be carefully coordinated to promote a thriving bio-economy.

The Amsterdam experience highlights some interesting challenges for looping organic waste. First, biomass and food waste are often mentioned together, but possess very different value chains. Biomass has greater potential value. It is less regulated, which means it is easier to loop. Food waste is more regulated and higher value products (from food waste) are relatively expensive. Thus, they are not currently competitive. Second, the organic waste chain in Amsterdam is very fragmented. There are many initiatives (from bio-refining, to composting, to reusing food waste), but they are not necessarily connected to one another. A clearer understanding of how circularity can be achieved in the chain and who is responsible for this is needed. The logistical organisation of the biomass and food chain remains problematic (e.g. the number of high-value flows is too small) and prevents high-value recycling. Thus, a regional strategy for high-value reuse of biomass with a special focus on smart logistical connections will be needed. This will have implications for the way in which the city-region develops.

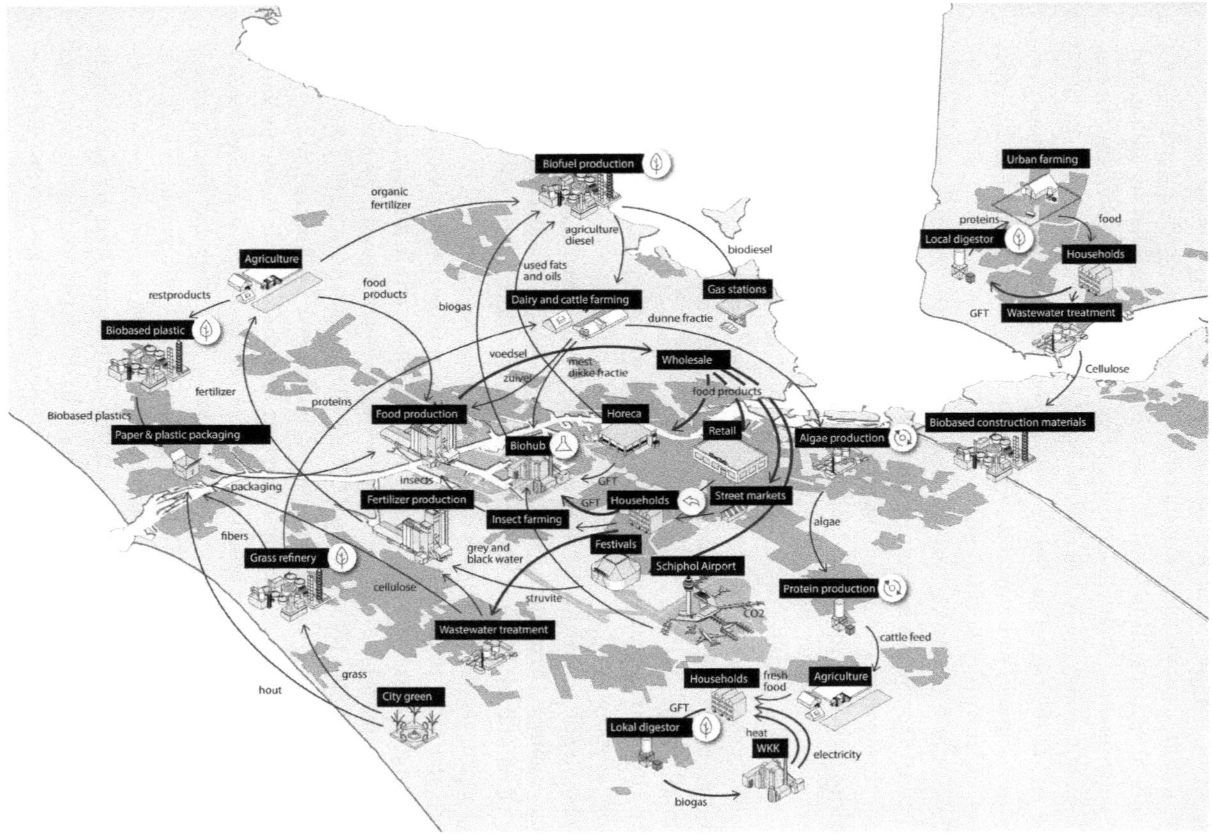

FIGURE 5.3 Vision of a circular organic residual stream for Amsterdam region.
Source: Bastein et al. (2016).

Circular De Ceuvel

De Ceuvel adopts another circular development pathway, which is temporary and experimental. De Ceuvel is found in the district of Buiksloterham. It is a living lab for circular development in Amsterdam's port area. Situated in the Noord district it is an example of post-industrial, waterfront reuse. Buiksloterham was originally designated an industrial area, with various activities including shipyards, petrochemical industries, the Dutch plane factory of Fokker, the Amsterdam incinerator and the northern power-plant of Amsterdam (Dembski, 2013). By the end of the twentieth century, the shipbuilding industry had disappeared and other industries had moved to Westpoort. With the exception of one repair shipyard, all the major docks closed.

The former NDSM shipyard was discovered by squatters and artists during the early 1990s who made it into a cultural hotspot. In the beginning this was without the assistance from the municipality. Later the municipality co-opted the cultural services provided in NDSM to regenerate the site. By the mid-1990s, the area was a mixture of large brownfield sites, active industrial sites, public utilities, workshops and new small-scale businesses and creative industries. However, 80% of the area was affected by soil pollution (metals and asbestos, volatile organic chlorine compounds and mineral oil), making it expensive to rehabilitate (Dembski, 2013).

The current regeneration of Buiksloterham focusses on the transformation of an industrial estate into a mixed-use urban neighbourhood. The aim of the redevelopment is to provide housing and a good living environment, while conserving established firms in the area. The plans envisage a gradual transformation of the economic structure of the area, from its traditional industrial base into a mix of green, creative and nautical industries. Buiksloterham will form the link between the more traditional urban development project (Overhoeks) and the new cultural district NDSM with its industrial character. It has been framed as a sustainable area for creative entrepreneurs and adventurous city-dwellers (Bosman, 2011). The plan is to develop 4,700 dwellings and create 8,000 jobs.

Buiksloterham is to be a living test bed and catalyst for Amsterdam's broader transition to becoming a circular, smart and bio-based city. It has many empty plots and almost no historic buildings. This creates space and flexibility for new development. However, it has scattered property ownership and many plots are highly polluted, creating prohibitive cost barriers to development. There is no masterplan for the area. It is the intention that the development will grow organically. The most important guidelines in the planning process are provided by the informal rules-of-the-game map. This is based on the set of goals presented in the *Manifest Circulair Buiksloterham*. These provide a vision for the site to be delivered by 2034.

One goal of the manifest is to close energy, water and nutrient flows. Another is to transition to a bio-based economy through the reuse of biological waste streams (e.g., nutrient recovery from organic wastes) and the use of bioprocessing to replace conventional industrial functions (e.g. soil phytoremediation instead of standard mechanical-chemical cleansing). Urban biodiversity and climate adaptation measures are also goals to bring long-term local resilience to the area. Finally, the key smart objective is to maximise social and environmental capital in the competitiveness of the area, through the use of modern infrastructure, highly efficient resource management and active citizen participation. Thus, the *Manifest* provides a framework for circular development. Key actors (Metabolic, Municipality of Amsterdam, several real-estate developers, Waternet and many others) signed the *Manifest*. As a consequence of this covenant, many circular development projects are being realised in the area including self-build projects, sustainable living on the water Schoonschip, PEK Ecostroom and Waternet's bio-refinery and De Ceuvel.

De Ceuvel is an excellent example of the temporary, experimental circular development pathway. It was developed on a former shipyard adjacent to the Johan van Hasselt kanaal. In 2012, the land was secured for a 10-year lease from the Municipality of Amsterdam. A group of architects (with a limited budget) won a tender to turn the site into a "regenerative urban oasis". The former industrial plot now provides a temporary home for a community of entrepreneurs and artists. It comprises creative workspaces, a sustainable café and spaces to rent. The neighbourhood has been designated as a "living lab". Thus, it is a test-bed for new circular technologies and for promoting circular lifestyles and practices (Figure 5.4).

De Ceuvel was built largely from recycled materials. Old houseboats were upcycled into energy-efficient workspaces, using recycled materials sourced from across the Netherlands. Upcycling is an important social practice on site as well as producing the visual aesthetic of De Ceuvel (loop). The water system was designed to close resource loops. For example, the dry composting toilets, separated urine collectors and struvite reactors recovered nitrogen and phosphorous to produce fertilisers (loop). Meanwhile, decentralised helophyte filtration systems[6] were used to recycle grey-water from the kitchen sinks (loop). The community also has an aquaponics greenhouse which recycles nutrients. The greenhouse produces vegetables and herbs for the local café. It uses a closed-loop aquaponics system combining fish and vegetable production. The fish excretes are broken down into nutrients for the plants. The plants provide a natural filter for the water. Inputs include primarily local nutrients like worms from composting bins and struvite from the urinals.

Waste heat is also reused in De Ceuvel (loop). Each office boat has a heat pump and an air-to-air heat exchange ventilation system. As warm air leaves the boat, over 60% of the heat is captured and re-circulated (Metabolic, 2013). The heat pump extracts heat from the surrounding air to warm each boat. These simple technologies remove the need for a gas connection and enable the use of renewable electricity to power the heating needs of each boat. This is particularly important on a contaminated site, where laying pipework would be risky. Over 150 photovoltaic (PV) panels are installed on the office boats. The panels

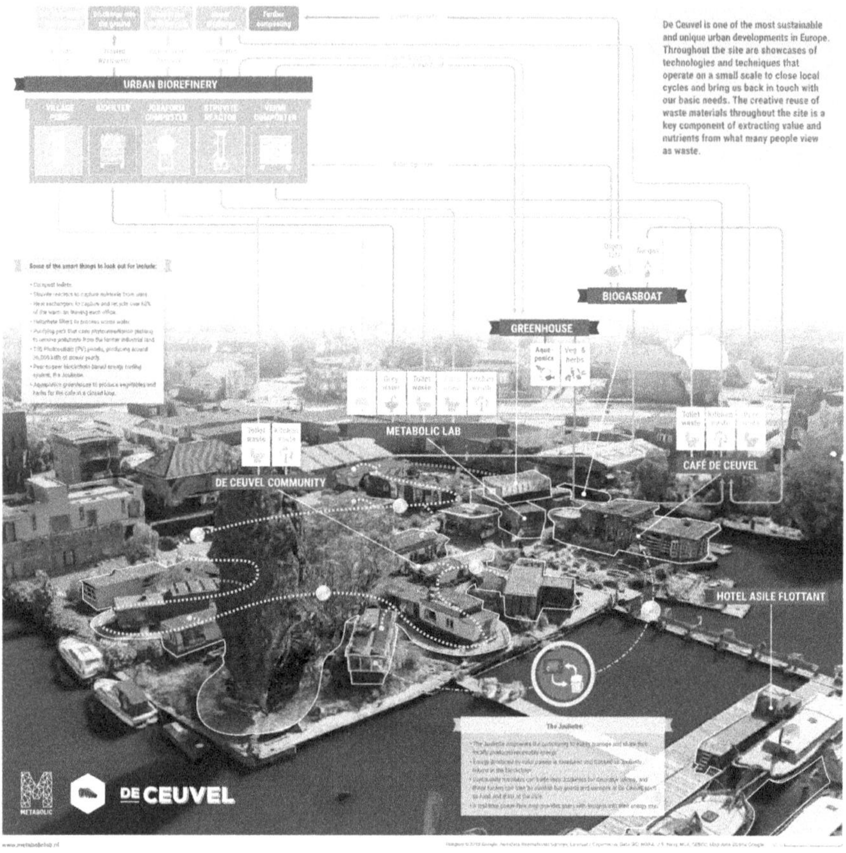

FIGURE 5.4 Looping systems in De Ceuvel.
Source: De Ceuvel website https://deceuvel.nl/en/about/sustainable-technology/

produce around 36,000 kWh of power yearly (ibid). This covers the electricity demand of the heating systems of the offices, along with a part of their residual electricity needs. The rest of the sites' power is supplied by a green energy company. A new crypto-currency was introduced in De Ceuvel, called the Jouliette. Those producing surplus energy are rewarded for production in jouliettes. This is facilitated by smart meters and block-chain technology. It encourages the local exchange of energy, instead of selling surplus power to the grid. The goal is to connect all of the Buiksloterham neighbourhood to create a local smart grid.

De Ceuvel was built on a heavily contaminated site. The houseboats were linked above ground by wooden walkways. Energy and water services were constructed above ground and off-grid to remove the problems of building subterranean structures in heavily contaminated soil. Phytoremediating plants were used to decontaminate the soil (ecological regeneration). The Municipality relaxed the planning conditions on the site, which would have prevented the use of heavily contaminated land without prior soil remediation. Instead, it required that the site was "safe, clean and healthy". This performance-based approach resulted in the relaxation of regulations, which helped to deliver circular innovation in De Ceuvel. This combination of phytoremediation and a performance-based approach offers an excellent strategy for urban ecological regeneration and reuse of vacant, contaminated land in post-industrial landscapes (adapt).

De Ceuvel has made the principles and strategies for circular development pathways, concrete and sharable. Such aesthetic materialisation is perceived positively by inhabitants and visitors. One key to success appears to be the designers' focus on inhabitants' practical requirements in conjunction with circular thinking. Inhabitant participation in the design process helped raise awareness of circular flows and the different systems used to manage them. Thus, the socio-technical system co-evolved through a co-design process. For a circular development to be successful, changing attitudes and social practices alongside new systems of provision is essential. The inhabitants affirmed that changing their practices in the neighbourhood was easy and quickly normalised. Project appropriation by the inhabitants was the basis for success. The inhabitants identified strongly with their neighbourhood and were proud of the living environment they had created. In addition, their emotional relationship with the living environment led to a virtuous loop of good practice in terms of resource management. Thus, these processes will also have increased the community's adaptiveness.

The temporary nature of the project is also important, since it enables the municipality to be more flexible in terms of the regulatory controls placed on the experimental development. However, the paradox is that this makes De Ceuvel as a model for circular development, more difficult to replicate in other urban locations, particularly because of political and economic barriers. Whilst these experimental projects are seen only as temporary uses, regulations can be relaxed and lower-value activities can be allowed. However, for long-term uses the regulatory requirements are greater and higher-value activities are expected. Thus, the De Ceuvel experiment might be replicated and could scale-up, but only if there were significant systemic changes within the development regime.

De Ceuvel demonstrates a circular development pathway for post-industrial sites. It shows how the remediation of contaminated brownfield sites can be achieved by raising structures above ground; using phytoremediation to decontaminate the soil and applying off-grid water and energy systems which do not require ground-works. The temporary nature of the project also helps to make it politically and economically acceptable, by indicating that it is a stage in a regenerative process, rather than an end in itself. Yet the temporary nature of the project undermines its ability to transform the wider development regime.

Summary

Amsterdam does have a circular strategy, which is clearly linked to sustainable development. Motivations for adopting this strategy are largely economic and environmental. Amsterdam recognises the difference between circular economy and circular development. It has programmes in place to address both. It also demonstrates two distinct pathways for circular development. The first pathway encourages a strategic, city-regional approach to looping construction and organic waste. The second pathway adopts a grass-root, temporary, experimental approach to circular development. De Ceuvel demonstrates how circular land issue and tendering can be used effectively to encourage circular development, particularly where conflicting restrictions are relaxed and a performance-based approach to development is taken. Amsterdam also shows how important it is to have a city-regional approach to planning development, in order to support circular resource flows.

Notes

1 Its ecological footprint (5 gha/capita) is considerably greater than its national biocapacity (0.8 gha/capita) which is limited.
2 Apart from biomass, generally available raw materials are the raw materials that nature needs for life (iron, silicon, carbon, magnesium, sodium, potassium, calcium, nitrogen, oxygen, phosphorus, sulphur, hydrogen).
3 The project incorporates an existing building in the new office complex, whilst reusing construction materials from the old building and creating flexible, dismountable, reusable upper floors.
4 A start-up from the University of Amsterdam, who extracts limonenes and fibres from citrus peel.
5 A company building a demonstration-plant for fermentation of organic waste into fatty acids.
6 Helophyte filters are simple constructions built using different layers. Sand, gravel and shells help remove solids, and a mix of special plants consumes organic matters such as nitrogen and phosphorus. Once purified, clean water is then discharged into the ground.

6

CIRCULAR PARIS

Circular economy in France

In France, the national government provided strong institutional support for the circular economy. By 2018, it had published a national strategy (Ministry for an Ecological and Solidary Transition and Ministry of Economy and Finance, 2018). The strategy focussed on increasing the efficient use of resources in production systems and supply chains. The design, supply, production and distribution of goods, materials, services and to an extent infrastructure are included in the strategy. It focuses on organic, plastic and construction waste streams. The ultimate aim is to reduce waste going to landfill. Other resources – water, energy and land – are not mentioned. The strategy does support industrial and territorial ecology (a closing of resource loops in industrial parks, at a city-regional and regional scale). Thus, it acknowledges the importance of the spatial, scalar and sectoral dimension of circular processes.

There are three key motivations for the circular transformation in France. First, moving to a circular economy would assist in reducing greenhouse gas emissions and delivering the United Nation's Sustainable Development Goals (SDGs). Paris hosted the 2015 Climate Action talks which produced the Paris Climate Agreement. The climate commitment underpins the implementation of the SDGs. France is keen to show its leadership in delivering both.

Second, after the recession of 2008, France wanted to reinforce its economic recovery and growth. It was estimated that the circular economy could create 300,000 additional jobs nationally and broaden the economic base (Ministry for an Ecological and Solidary Transition and Ministry of Economy and Finance, 2018). Thus circular economy could generate growth. France also wished to reduce its dependence on imports of raw materials and its vulnerability to global economic uncertainties. Through the reuse, recycling of resources and energy recovery France could begin to reduce its imports of goods and oil.

Third, the French government was also keen to address issues of social injustice and inequality. The circular economy potentially offered an opportunity to do this. Circular economy could be a source of social innovation and could offer a range of employment opportunities to the most vulnerable. It would enable the entrance of new players into the market (including social enterprises) who prioritised societal goals. It could provide a vehicle for the redistribution of resources, which would help to deliver social solidarity goals. The French government also thought it was important to demonstrate its commitment to inclusivity in the process of creating the national circular strategy. Thus, a wide range of actors (including social enterprises and community groups) were engaged in the process.

Circular Paris

Paris's population is predicted to remain stable at 2.23 million until the middle of the century (French National Statistics Agency, 2020). However, the ecological footprint for the average Parisian (4.8 gha/capita in the Ile de France) is growing. Food accounts for 50% and consumer goods 30% of the region's footprint (IAU, 2016). Currently, the footprint of those living in the Ile de France is nine times greater than the region's bio-capacity. Thus, taking a circular approach to development will be crucial for Paris's future resilience.

There were a number of motivations for moving towards a circular economy in Paris. The first was Paris's commitment to deliver zero waste, particularly to tackle organic and construction waste in the city-region. Paris committed to the zero waste target in 2014, which was reinforced by winning a national call in 2015, for projects to create a zero waste territory. The Paris Urban Planning Agency (IAU) suggested circular economy could deliver the zero-waste goal. The adoption of circular economy principles in Paris was further supported by the national "Energy Transition for Green Growth" law, which was adopted in 2015. This sought to reduce waste going to landfill, largely by increasing reuse, recycling and energy recovery.

The second motivation was to become sufficient, through the local production and assimilation of resources. This led to the introduction of programmes to boost local agriculture and re-industrialise the region. Paris is highly dependent on its immediate and wider environment for resources. The region imports the equivalent of 12 tons and exports 5.2 tons of materials per resident per year (IAU, 2016). Paris intends to close resource loops within the city-region where possible by localising construction, food and industrial systems and encouraging looping actions.

The third motivation, Post-COP15, was to reduce greenhouse gas emissions. Circular actions were identified as critical to mitigating climate change. The climate action plan suggests a range of circular actions – energy and heat recovery from waste; renovation and energetic refurbishment of building stock; localisation of resource flows; reuse of goods; recycling of grey-water and wastewater – for tackling greenhouse gas emissions. It also highlights the importance of the restoration of urban ecosystem services, to sequester CO_2 and enable the city to adapt to climate change.

The fourth motivation was social solidarity. The aim was to redistribute "residual" resources (e.g. food, furniture, electronic goods) to those in need in the city-region, to establish community solidarity projects and social enterprises to support the process, to engage and empower excluded groups through the creation of new employment opportunities and thus to build resilience within communities. Also through temporary interventions to enliven, re-invent and transform urban spaces with citizens.

Paris adopted a plan for circular economy in 2017 (Figure 6.1). It produced a more holistic and territorially embedded view of circularity. Materials, water, energy, land and infrastructure were considered in the plan. Linkages between looping, adaptive and ecologically regenerative actions were also recognised. This approach more clearly reflected circular development defined in Chapter 2. The plan was produced through a highly inclusive consultation process conducted by the General Assembly. In total, 20 local authorities and more than 120 organisations (from the non-profit, industrial, government and academic sectors) took part in its creation (Marie de Paris, 2017).

The city has used several levers to implement the plan: incentive-based funding, public procurement, regulation, data platforms, skills workshops and experiments. The main source of public funding was through public procurement, which provided powerful leverage for a circular transition (Marie de Paris, 2017). Paris created a trans-national procurement group with several other European cities in 2015. It was the first to have its responsible public procurement scheme approved. One of the main goals of the scheme was to support the growth of the circular economy. Paris defined new resource efficiency criteria (both material and human) for its future public procurements, which would help deliver the most resource-efficient solutions for the region.

Paris is building capacity for implementing the circular economy. The highly inclusive, multi-stakeholder process, organised to develop the plan, began the capacity building process. It helped to create the social networks and reciprocal relationships between actors in the city required to encourage circular solutions. The municipality also set-up data platforms (e.g. *Paris Urban Metabolism Platform*) to enable actor learning and to identify opportunities for resource exchange (e.g. *sol-dating*).

A re-industrialisation strategy for Paris was introduced, with circular economy as a key focus. Networks of economic actors (such as the *Paris Esprit d'Entreprise*, the business and employment centres, and the *Groupement Jeunes Créateurs Parisiens*) were established to enable the process (Marie de Paris, 2017). These projects were supported financially, technically and though property mobilisation (similar to London) to encourage their development and scaling-up. The initiative "Paris City of Makers" aimed to double the number of production spaces in the capital (fab labs, makerspaces, etc.) which could further reinforce the re-industrialisation process and enable new circular industries to emerge

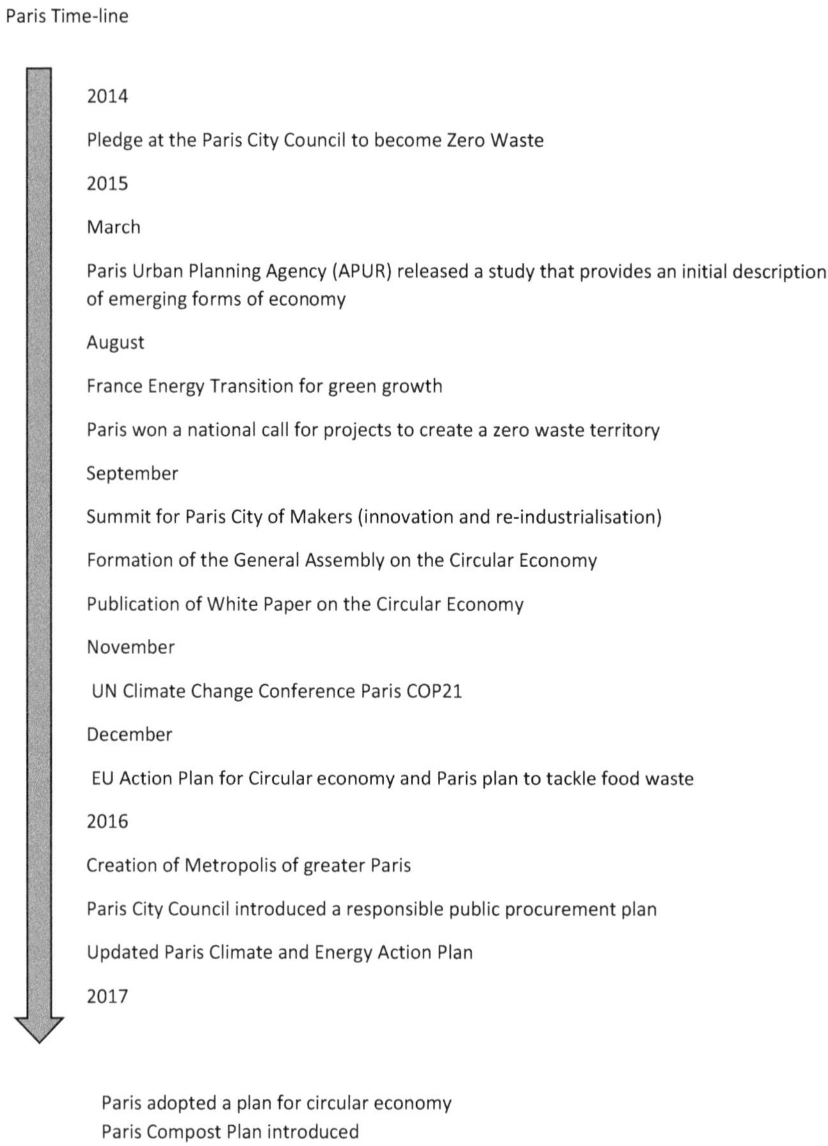

FIGURE 6.1 Time-line.
Source: Author's own.

(Marie de Paris, 2017). Over time it will broaden the economic base of Paris, producing new employment opportunities, resulting in economic growth.

Experiments make it possible to explore the challenges to the implementation of the circular economy (in terms of the legal framework, economic model, networking of actors) in Paris. The *Improvement of the Urban Metabolism Programme* has produced a range of experimental projects focused on the reuse of property, materials or excavated soil; the collection of organic waste and raising awareness of wastage through education campaigns (ibid). The lessons learnt from these experiments have been disseminated through the innovation platform.

Circular development in Paris

The circular economy plan also provided a territorial context for the resource flows (i.e. the metropolitan region). It recognised the link between land-use (activities), the production and consumption of resources. Thus, it recognised the link between spatial planning and the circular economy. The main thrust of the circular economy plan was towards material reuse, recycling and energy recovery. It prioritised looping organic, food and construction waste. It also tackled plastic waste and waste goods (particularly electronic goods and furniture). It highlighted the importance of shortening production and distribution chains, whilst ensuring the visibility of the good supply chains. Strategies to close resource loops locally were prioritised.

Due to the large number of construction sites in Paris, vast quantities of materials are imported. Yet construction waste continues to grow. Construction waste represents 45% of materials exported from the region (Marie de Paris, 2017). This is a particular problem in the Ile-de-France which generates 30 million tons of construction waste per year. This is more than five times the production of household waste for the area. Of these 30 million tons, less than 25% is currently recycled (ibid). Reducing construction waste by encouraging the adaptive reuse of buildings and the local recycling of materials and components is central to circular development. It can also be encouraged through the planning process (as was the case in Amsterdam).

The municipality also seeks to enable the creation of local food systems by protecting farmland within the city-region. This creates a closer relationship between producers and consumers, which in turn raises awareness of the impact of consumption choices locally. Farming can help to protect ecosystem services and increase local sufficiency. Organic waste produced in the city can be composted and used as fertiliser, or converted to energy in the periphery (in bio-digesters). Many local authorities in the region still own significant parcels of farmland. However, these parcels are threatened by urbanisation. The regional plan (Regional Authority of Greater Paris, 2013) includes a protection and development objective for farmland. There are also plans to encourage urban agriculture and composting within the city-limits. Closing food loops locally offers a key circular development pathway, which can be supported by spatial planning.

Paris is attempting to re-industrialise the city-region. The localisation of industrial systems will increase resource sufficiency and reduce greenhouse gas emissions from the transport of goods and materials. The formation of regional industrial clusters may also enable resources to be shared, reused and recycled locally. The Île-de-France is a potential producer of many secondary raw materials, some of which are already "rare" and essential for the development of future industrial sectors. If all waste produced in the territory was recovered, 9% of ongoing needs for raw materials within Île-de-France could be met (Marie de Paris, 2017). Île-de-France also has many assets to enable this transformation. It has a vast network of companies in key areas for the circular economy, including mature recycling schemes (for metals, paper, etc.), and developmental actors in the chemicals, energy and design industries. Networks for industrial symbiosis are more likely to form where industry is present and clustered. Spatial planning has a key role in encouraging the development of these clusters.

In Paris, the spatial plan determines infrastructural provision, urban form and land-use activities which directly impact on a city's ability to loop resources locally. This is reinforced by an amendment to the plan which uses regulatory provisions to prioritise resource conservation and environmental quality. Thus, Paris has produced an extensive network of voluntary waste drop-off sites linked with recycling centres. It has strengthened quality requirements and the standards governing green infrastructure, which improves rainwater management. It has led to the inclusion of decentralised, renewable energy systems in new development. Through its powers of provision, the municipality is also responsible for providing water, sanitation and waste management infrastructure and services. Thus, Paris can adopt new infrastructural systems or models of provision which could support grey-water recycling systems, anaerobic digestion of organic waste, waste-heat capture, recycleries, logistics centres and storage facilities.

Paris takes an integrated approach to circular development, where synergies between circular economy (looping activities), nature-based solutions (ecological regeneration) and climate adaptation are acknowledged (Table 6.1). Several plans encourage looping, regenerative and adaptive actions to operate together (e.g. local urban plan, the biodiversity plan). Equally, a transition towards a circular economy is promoted through many strategic plans (e.g. air quality plan, climate and energy plan). The recognition of these important interactions by government is materially different from the approach taken in London, where these actions are viewed separately. This integrated approach is operationalised strategically across the city and manifests in new eco-districts (e.g. Clichy Batignolles).

Land release is an important instrument for circular development. The *Paris Reinvented, Parisculteurs* and eco-district programmes help to deliver circular development. These programmes all rely on the allocation of sites in Paris. Land (publicly and privately owned) is released for projects which are innovative, conserve resources and improve environmental quality (e.g. Milles Arbres, Agripolis and Clichy Batignolles). Temporary planning permissions are also a powerful tool for circular development. Privately owned sites are leased (usually by land owners or developers) on a temporary basis for a range of pop-up circular activities (e.g. Les Grands Voisins, Bellastock). Three circular development pathways have emerged in Paris: city-regional (with a focus on construction and food waste, water and vacant sites), temporary-experimental (with focus on vacant sites) and eco-district pathways. All three will be illustrated below.

TABLE 6.1 Inter-relationships between plans promoting circular development

	Circular economy	*Ecological regeneration*			*Adaptation-resilience*		
	Promote transition to CE	*Improve environmental quality*	*Promote biodiversity*	*Mitigate and adapt to climate change*	*People centred resilience*	*Urban form and infrastructure centred resilience*	*Mobilising collective intelligence*
Climate & energy plan	λ	λ		λ	λ		λ
Local urban plan	λ	λ	λ	λ		λ	λ
Air quality plan	λ	λ		λ		λ	λ
Noise prevention plan	λ	λ				λ	λ
Environmental health plan	λ	λ			λ	λ	λ
sustainable urban logistics	λ					λ	
Biodiversity plan	λ	λ	λ	λ		λ	λ
Blueprint for the ecological restoration of waterways	λ		λ	λ			
Paris rain-plan	λ		λ	λ		λ	λ
Blue-print for recovery of non-potable water	λ		λ	λ			
Urban agriculture development strategy	λ	λ	λ	λ		λ	λ
Sustainable Paris actors network	λ	λ	λ	λ			

(*Continued*)

	Circular economy	*Ecological regeneration*			*Adaptation-resilience*		
	Promote transition to CE	*Improve environmental quality*	*Promote biodiversity*	*Mitigate and adapt to climate change*	*People centred resilience*	*Urban form and infrastructure centred resilience*	*Mobilising collective intelligence*
Local housing programme	λ	λ			λ	λ	λ
Sustainable food plan	λ			λ	λ	λ	λ
Plan urban agriculture and food	λ			λ			
Zero-waste plan	λ	λ		λ		λ	λ
Compost plan	λ	λ		λ			
Plan food waste	λ			λ	λ		λ
Innovation arc	λ						
Responsible public procurement	λ					λ	λ
Participatory budget	λ	λ	λ	λ		λ	λ
Local waste and recycling plan	λ	λ			λ	λ	λ
Circular economy roadmap	λ				λ	λ	λ
Flood prevention programme		λ		λ	λ	λ	λ
Smart and sustainable city	λ	λ	λ	λ	λ	λ	λ

Source: Adapted from Marie de Paris (2017).

Circular construction (looping actions)

Paris's approach to construction waste demonstrates the city-regional, circular development pathway. Construction and renovation projects generate large flows of materials which have an impact on greenhouse gas emissions and air pollution. To minimise the environmental impact, the City of Paris is encouraging renovation. It also requires that all construction projects adopt circular economy principles in order to attain the target of 50% construction waste diverted from landfill by 2030 and 100% by 2050 (ibid). The use of materials with lower embodied carbon, such as locally produced wood, or recycled materials from sites in the city-region, is also encouraged.

Deconstruction sites are emerging in Paris, which enable the recovery of construction materials from refurbishment projects and demolition. Two logistics centres have been funded on the periphery of Paris, to enable materials to be stored and re-distributed in the region. Through contracting arrangements, public works are beginning to adopt circular construction practices and use locally sourced, recycled components or materials. Intelligence is being gathered to establish local supply chains. Various events have been organised to stimulate regional networking, to optimise the reuse and recovery of materials at the local level.

Experimental projects are also emerging to determine how locally sourced materials can be reused in construction projects. A particularly innovative experiment is being conducted by the students from the Architectural School of Belleville. They have developed a festival and research platform called Bellastock, which aims to test innovative methods for reusing construction waste. Bellastock holds an annual construction festival, which promotes experimentation in architecture (adaptable buildings) and construction material reuse. Architecture students from across France build an ephemeral city with salvaged materials (e.g. pallets, plastic and earth). This experience helps the students (future architects) to understand the challenges to construction and operation.

Bellastock also run the Actlab experiment. This is a laboratory for the reuse of construction waste, located on the construction site of the Ile-Saint-Denis eco-district. The site is currently undergoing a transformation, as part of the Olympic development for 2024. The former industrial uses are being replaced by residential and commercial uses. The structures which were on site, including a warehouse for Printemps, have been deconstructed. Actlab has used the components and materials collected from the warehouse to experiment. The Actlab is complemented by the REPAR project (also developed by Bellastock) which aims to produce a guide for integrating reused and recycled materials into architectural projects and a catalogue providing technical solutions.

Soil excavated from building sites is another issue in Paris. A soil exchange service – SOLDating – launched in 2014, acts as a broker for the supply and demand of inert soil between project sites. It facilitates soil traceability, and thus enables soil exchanges between sites via a simplified platform. The platform recovers and reuses building and public works soil that is not contaminated. It has already recovered more than 30,000 m^3 of soil in the Île-de-France region (Marie de Paris, 2017). It is estimated that decreasing these flows could result in a 50% reduction in inert soil management costs (ibid). The establishment of a mixing platform to build fertile soils (from demolition waste, cleaning sludge and green waste) further limits the volumes of inert site waste and sludge discharged into storage centres. In the long term, the proposed fertile substrate may be an alternative to an increasingly distant supply of topsoil.

Circular food (looping, ecologically regenerative and adaptive actions)

Paris adopts a city-regional approach to creating circular food systems. The aim is to grow, reuse, compost and recover energy from food-waste, within the city-region. Urban agriculture (in all its forms) has been present in Paris for more than 150 years. "La culture maraîchère" (market gardening) accelerated in Paris during the second half of the nineteenth century and peaked during occupation in World War II. Urban agriculture was carried out in collective workers and family gardens on leased plots. The tenure of these gardens was secured by national law. The value of urban farming to increase food security, improve Parisian's health and enhance the urban ecosystem was understood by the government.

Post-industrial landscapes in the Parisian suburbs afforded plenty of opportunities for urban agriculture (Demailly and Darly, 2017). The expansion of family gardens followed the gentrification process into the suburbs. Today, Paris is an extremely dense city. Only 17% of the land area is green space (ibid). It is in this context that temporary forms of urban farming have emerged and now contribute to the diversity of institutionalised gardening and guerrilla gardening projects in vacant spaces throughout the region. Since 2000, the sites available for agriculture have become smaller, more dispersed and ephemeral (ibid). Thus, increasingly urban agriculture is becoming a temporary (rather than sustained) activity.

Several new forms of urban agriculture have emerged: permanent collective gardens and farms, temporary gardens, nomadic animal herding and off-the ground gardens. Unsurprisingly, most urban agriculture initiatives can be found in interstitial spaces left vacant by the urban economy (ibid). If urban agriculture of this kind is not to be threatened by land scarcity (and the low-value nature of this activity) it needs to be protected. Alternatively, the relocation of agricultural projects needs to be planned more systematically.

In response to this, the Parisian government is supporting urban agriculture through the "Parisculteurs" initiative. The goal is to cover the city's roofs and walls with 100 hectares of vegetation by 2020 (formalised by the "100 hectares charter" in 2016). One third of this space will be dedicated to urban farming. Land owners and managers of public and private spaces (currently a 30:70 split) can nominate their space for conversion to agriculture. A web platform is used to link farmers with space owners/managers. Often, financing for projects comes from crowd-funding, for which the returns are good. However, in practice, farming in and on buildings is still almost non-existent in Paris. It is expensive in terms of capital and operational costs. Nevertheless, 33 "greening" projects in and on buildings have been supported and are now in the early stages of implementation.

One such project is the largest roof-top, urban farm in Europe (Figure 6.2). It is being constructed in the 15th arrondissement. The farm will cover 14,000 m^2 once complete. It is located on the top of a major exhibition complex (Paris Expo Porte de Versailles). The site will produce 1,000 kg of fruit and vegetables every day in high season, using entirely organic methods (Harrap, 2019). The project is being developed by Agripolis (an urban farming company) as a commercial venture, which will protect it from the constant competition for space in the city.

Food waste presents a major problem for Paris. Parisians discard three times as much food still in its packaging compared to the average French citizen. Despite efforts to reduce food waste, around 37% of the food served is finally disposed of as bio-waste, most of which is being incinerated (Marie de Paris, 2017). The City's objective is to reduce Paris's greenhouse gas emissions from food by 40% by 2030.

Food reuse is one strategy adopted in Paris. This approach tackles the food waste problem, whilst building social solidarity, by feeding the urban poor. Food reuse schemes have been reinforced by the law introduced in 2016 (Perchard, 2016), which made it illegal for supermarkets to dispose of good quality food in

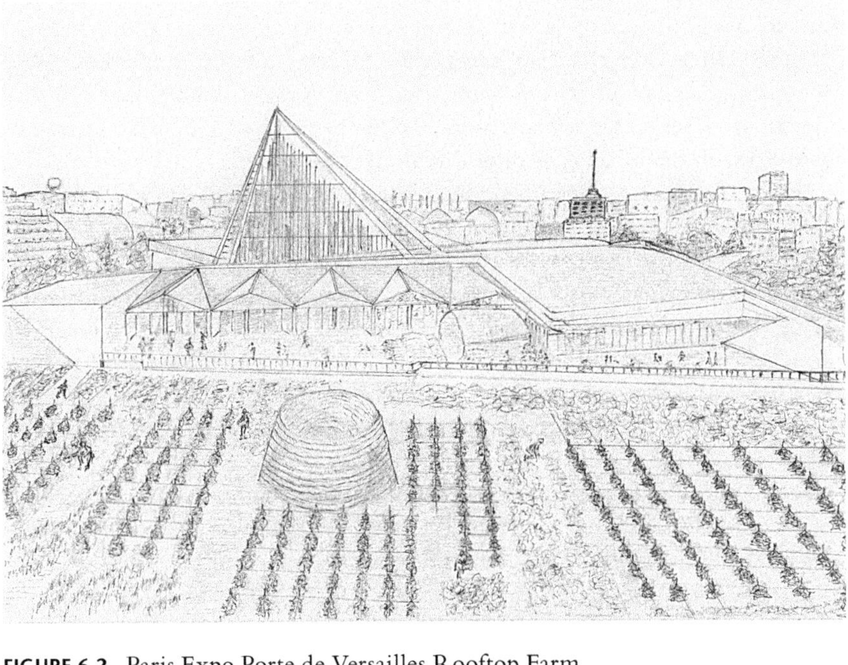

FIGURE 6.2 Paris Expo Porte de Versailles Rooftop Farm.
Source: Artistic impression produced by Sally Williams.

France. The law forced supermarkets to seek alternatives to landfill. The City of Paris is providing some support to non-profit organisations for the collection of unsold food items from commercial enterprises including supermarkets.

As part of new public service contracts for the management of Paris's food markets (2015–2019), contractors have been required to develop partnerships with local non-profit organisations in order to organise the redistribution of unsold edible fruit and vegetables (Marie de Paris, 2017). Ten markets are now part of the scheme. One recipient is the non-profit Freegan Pony restaurant which reuses discarded food from the wholesale market. Other recipients are the 50 to 80 gleaners in the Joinville market (19th district) who collect 300–400 kilos of unsold food every Sunday (ibid).

Some non-profit organisations have received a subsidy to equip themselves with logistical means or kitchen equipment in order to recover unsold food from supermarkets and prepare meals for those most in need. Another innovative, solidarity project enabling food reuse is Les Frigos Solidaires. These are fridges placed in the community where businesses and households can deposit surplus food for those who need it. The fridges are located in or close to local businesses which maintain them. Paris gave a grant in 2018 to finance 15 fridges across the city.

Composting food waste offers another option for closing the food-loop locally. The City of Paris encourages collective composting in all its forms. However, it is difficult in a city with many high-rise buildings. Nevertheless, there are 422 household composting sites at the foot of buildings and six neighbourhood compost bins and there is still significant growth potential (Marie de Paris, 2017). The Paris Compost Plan, adopted in 2017, sets a target of 500 household composting sites in collective housing (twice the current number) and 400 sites in public facilities (twice as many as in 2016) to be met by 2020 (ibid). These can be linked with the urban agricultural projects across Paris.

Finally, food waste (unsuitable for reuse or composting) can be converted to energy. Paris currently imports 95% of the energy it consumes. Organic waste provides the opportunity for energy recovery, reducing energy imports and increasing energy security in Paris (ibid). Since 2014, Paris has been deploying solutions to generalise the collection of food waste produced by municipal canteens, schools and food markets. The food bio-waste is then valorised in an anaerobic digestion plant in Paris, to generate biogas, which is reinjected into the city gas network or used as a bio-fertiliser for green spaces in the capital (ibid).

The City has set itself the target of establishing medium-sized biogas production by 2030. It is also planning to develop organic waste recovery facilities (biogas production and industrial composting) in the metropolitan area, including the installation of a high-capacity biogas plant. This will enable Paris to process waste flows, whose volumes will inevitably increase with the obligation to implement separation at source by 2025 (imposed by the Law on Energy Transition for Green Growth).

Circular water (looping, ecologically regenerative and adaptive actions)

Paris takes a city-regional approach to the development of a circular water strategy. Water cycles can be assisted by looping and optimisation actions. Improving cycles increases water security in cities. In 2013, Paris used 176,000,000 m^3 of water annually (Tabuchi, Tassin and Blatrix). Consumption has reduced by 25% since 1985. The network is very efficient (only 5% leakages) and more recently the use of water metering has helped to reduce potable water consumption further. In 2010, water was re-municipalised in Paris. Thus, the city has direct control over the provision of both the service and infrastructure (potable and non-potable). This is useful in enabling its circular transformation.

Paris is one of the few cities in the world equipped with a dual network water system: the drinking water network is duplicated by a totally independent non-potable water network which possesses its own means of production, storage and distribution pipes (Nguyen, 2003). In this second network there circulates non-potable, grey-water. The grey-water system has existed in Paris for two centuries. Grey-water is very inexpensive and is used in large quantities. Most of the grey-water (98%) is consumed by the city of Paris and used for hydrants, fountains, street cleaning, watering of public gardens, flushing of the sewers, etc.

The grey-water network will require replacement in the near future, and there are some questions surrounding whether this is economically or ecologically prudent (Nguyen, 2003). Ecologically having one system would reduce the water lost to leakages. Also potable water is more expensive and metered, so moving to a solely potable water system should reduce overall water consumption. Economically, the cost of removing the grey-water system is extremely high, but it would release valuable land for development.

Paris is now investigating replacing existing grey-water reuse systems with blue-green infrastructure. These systems have other added advantages, for example, in adapting the city to climate change. Thus, more localised, blue-green, grey-water reuse systems are beginning to emerge in Paris in existing and new developments. The Climate Adaptation Plan (Marie de Paris, 2018) and the Blueprint for the Non-Potable Water (Marie de Paris, 2015) sought to upgrade blue-green infrastructure in Paris to cool the city, reduce water scarcity and adapt to climate change.

Rainwater gardens, permeable surfaces, swales, reed-beds and retention ponds are being used to create sustainable urban drainage systems across the city. This will help to regulate the retention and flow of water. Reflecting ponds, temporary swimming pools and cool pathways are being used to regulate the urban heat island effect. An additional 20,000 trees will be planted, 100 hectares of green roofs and walls and 30 hectares of green public space will be provided in Paris by 2020, which will enable grey-water storage and reuse.

The potential to capture heat from the waste-water and non-potable water network is also being explored (looping action). Heat is already being recovered from waste-water experiments in Paris. For example, the experiment in the Wattignies School which recovers 70% of its heating needs (reducing CO_2 emissions by 59%) from the waste-water system, by the combined use of a heat exchanger and heat-pump (Marie de Paris, 2017). Another example is the Chevaleret experiment. This is an innovative waste-water heat recovery facility coupled with a mini-gas cogeneration plant. It saves more than 50% of the energy cost compared to a traditional gas solution (Marie de Paris, 2017). A study highlighted a further 13 sites with energy recovery potential from the wastewater and non-potable water network in Paris. This could save up to 250 tons of CO_2 each year (Marie de Paris, 2017).

Separated urine management is also being tested in eco-neighbourhoods in Paris (similar technology to that used in De Ceuvel). The installation of urine collection systems at the source can reduce the flows of nitrogen-rich and phosphate-rich nutrients entering wastewater treatment plants and the River Seine, and cut water consumption by reducing the flushing of toilets. Therefore, the separation of waste-water at the source is also a promising solution in terms of adaptation to climate change and the responsible consumption of water resources. If separated urine collection is implemented on a large-scale and it is processed appropriately, the recovered nutrients could be used as fertiliser. However, these open-air installations – uritrottoir – have become rather controversial in Paris.

Circular urbanism (looping, ecologically regenerative and adaptive actions)

Circular urbanism is critical to the reinvention and adaptation of a city. It allows the reuse of "wasted" spaces (sites and buildings) through a temporary, experimental approach to development. It is particularly important for low-value circular activities, which would otherwise find it difficult to compete for space, in a land scarce city such as Paris. Increasingly land owners or developers are searching for temporary uses on sites. This ensures land is used effectively between longer-term uses. During this time temporary activities occupy the space. These provide the opportunity for looping, ecologically regenerative and adaptive actions.

Temporary uses have become increasingly popular in Paris since 2012 (Figure 6.3). Amongst these temporary uses, circular activities have emerged, for example, The Freegan Pony (food waste reuse café), Friche Miko (an entertainment venue on wasteland which also demonstrates circular economy), Jardin d' Alice (adaptive reuse of buildings for workshops and performance spaces, also with eco-projects in outdoor space) and Actlab (reuse of construction waste).

A particularly successful example of a temporary circular experiment is Les Grands Voisins, which occupied the site of Saint-Vincent-de-Paul hospital in the 14th arrondissement (Figure 6.4). The developer decided to offer the site to temporary uses, rent-free until development began. The hospital's buildings were turned into a hub of social and commercial enterprise. There was a hostel

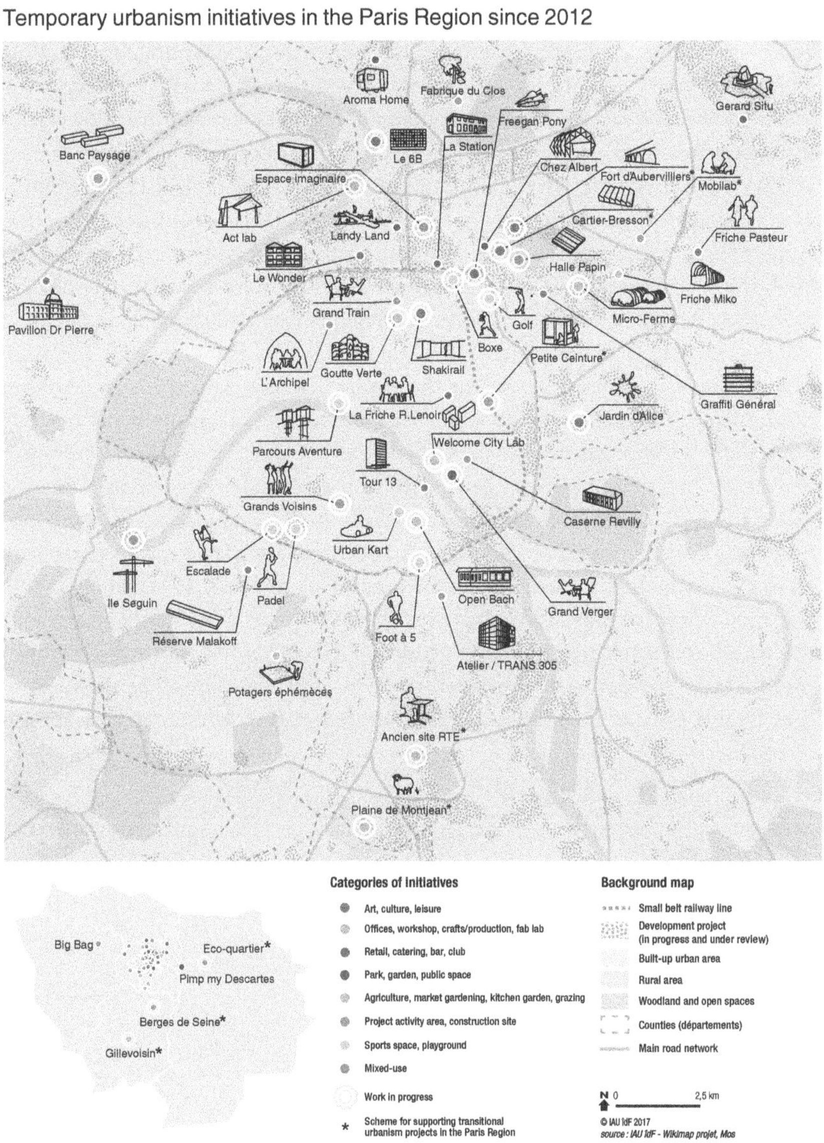

FIGURE 6.3 Temporary Urbanism initiatives in Paris post-2012.
Source: Adapted from IAU îdF, Lieux culturels éphémères. Promenade cartographique en Île-de-France.

providing 600 beds for the homeless, workshops for artisans, pop-up shops and start-ups. The former ambulance bays and car parks were converted into allotments, a boules court, a makeshift football pitch and an urban campsite. As many as 1,000 visitors came daily to its market or cafes or to see live performances. The aim of the project was to tackle social exclusion and encourage social solidarity.

Preparatory building works began in 2018 which threatened the temporary uses on site. The developer suggested they move to an abandoned area along the Avenue Denfert-Rochereau (adjacent to the site). This further extended the lifetime of the project through a secondary temporary occupation. Les Grands Voisins demonstrated an alternative development trajectory for Parisians. It illustrated how buildings and materials could be reused and recycled. It also demonstrated demand for some circular activities (e.g. sale of second-hand goods, repair and upcycling, urban agriculture and composting). It built social and human capital amongst communities using the site, which increases adaptive capacity within these networks. It also ecologically regenerated and reinvented the area.

Many temporary experiments have emerged across Paris since its inception in 2011. The key actors involved in the project (L'Association Plateau Urbain, Yes We Camp, Aurore) have also been engaged in the scaling-up of the practice. For example, L'Association Plateau Urbain has become a broker for temporary urbanism. It links organisations in the cultural or social sectors with owners of unoccupied buildings and sites to enable reuse. It was a small association before Les Grands Voisins project, but now it is an important player, often partnering with the municipality, to deliver sites for temporary uses. *Yes We Camp* is a social enterprise, whose purpose is to create meanwhile spaces which engage the local community and support vulnerable groups. This group has successfully implemented a variety experimental projects in Paris[1] and Marseille since the creation of Les Grand Voisins. Aurore is a charitable association which was founded in

FIGURE 6.4 Les Grands Voisins – meanwhile space in central Paris.
Source: Artistic impression produced by Sally Williams.

1871. It accommodates and supports vulnerable people. It also has many temporary projects across Paris.[2]

These examples inspired the *Paris Reinvented* initiatives, instigated by the Mayor's office in 2014. It formalised the process of the strategic adaptive reuse of sites and buildings in Paris. It is another example of the city-regional (strategic) approach to circular development. Disused public sites were leased or sold to developers and architects competed with innovative plans for their redevelopment. However, these were not temporary projects. They were transformative. Requirements were placed on the projects to offer flexible land-use, anticipate new lifestyles, use wasted spaces and engage with communities. This created more resilient urban form and built adaptive capacity amongst key actors.

The first 23 sites selected were diverse but were rarely in iconic locations. The city gained approximately €565 million through the sale or lease of the sites (Pilsudski and Koh, 2019). Many innovative projects were proposed, of which three – Morland,[3] Mille Arbres[4] and Massena[5] – clearly demonstrated examples of circular development. They incorporated looping, adaptive and ecologically regenerative actions. The success of the initiative encouraged Paris officials to embark on new projects: "Reinventer la Seine" with the City of Le Havre at the mouth of the Seine River; "Inventons la Metropole" with the Metropolis of Greater Paris and "Subterranean Secrets of Paris". Thus, Paris combined city-regional and temporary experimental pathways to enable the circular transformation of redundant urban spaces.

Circular Clichy-Batignolles

Clichy-Batignolles is an eco-district which materialises all the circular actions. It is constructed on a 54 hectare, contaminated, brown-field site. The site was the railway yard next to Saint Lazare train station in the 17th arrondissement (Marie de Paris, 2015b). The redevelopment of the site began in 2001. The intention was to create a village that would serve the 2012 Olympics. The bid was unsuccessful, but it led to the redevelopment of the site as an eco-district (loop). The district addresses the climate objectives identified in the Paris Climate Change Agreement. It is a demonstration project for a low carbon neighbourhood, which is also climate adapted (adapt).

Clichy-Batignolles provides an example of circular urban development. Infrastructure incorporated into the development enables grey-water recycling, waste-heat reuse, material waste recycling and energy recovery from waste (loop). The inclusion of the green infrastructure also helps to ecologically regenerate the urban system and adapt to climate change (regenerate/adapt). The neighbourhood when complete will provide accommodation for 7,500 residents and provide jobs for 12,000 people (Rougé, 2015).

Essentially, it comprises a 10-hectare park (Martin Luther King Park) surrounded by resource efficient, climate-adapted buildings (adapt). The buildings are extremely energy efficient (optimise). Energy consumption is restricted to

50 kwh/sqm/year (compared with the Paris building code which requires 70 kwh/m^2/year). For space heating passive house standard (15 kwh/m^2/year) has been set. Some buildings also have green roofs (covering 16,000 m^2) that offer insulation and a habitat for wildlife (regenerate).

Much of the district's heat and electricity comes from renewable sources (substitute). A local geothermal heating system taps into a warm water-table beneath the park, thereby reducing energy needed for heating. Nearly 4,000 tons of CO_2 are saved every year (compared to a heat network using natural gas). Some buildings also use heat from outgoing grey-water to heat incoming tap water (loop), saving 58% of the energy typically required for water heating (Marie de Paris, 2015b). More than 35,000 m^2 of solar panels have been installed on roofs and facades. These will generate 3,500 MWh per year, which is around 40% of the electricity used in the development (ibid). The layout of the development encourages walking and use of mass transit while limiting space for cars (substitute). To further improve air quality in the district, deliveries are restricted to a fleet of electric vehicles that cover the last kilometer from a central drop-off site.

At the heart of the district is the Martin Luther King Park, which is the largest green space in Paris's 17th arrondissement. The park has been designed to increase biodiversity, accessibility to green space and water management. It also acts as a cooling space within the urban environment (regenerate). The two community gardens on site also give residents places to grow their own food and compost food waste (adapt and loop). The park and other green infrastructure in the neighbourhood reduces the volume of rainwater entering the sewage system (loop/regenerate). It includes wetlands and subterranean tanks that collect rainwater for reuse (loop). This reduces the volume of waste-water being treated and contamination problems from overflow. It also captures rain, which can be used to water the green infrastructure in the district. Of the parks watering requirement 40% is met by the rainwater recycling systems.

The district's waste management system uses underground pneumatic tubes, for transporting waste to the sorting centres (like Hammarby). Recyclable materials are then transported to reprocessing plants by train. The remaining waste is compacted and taken to the Saint-Ouen incinerator. Only three to four trucks are needed to remove waste per week, making a saving of 1,872 km in transportation per year (ibid). This reduces greenhouse gas emissions (42%), carbon monoxide (98%), nitrogen oxide (86%) and particulate emissions (90%), when compared to a typical waste collection system (ibid).

Summary

Paris has a circular strategy, which offers a more holistic and integrated conceptualisation of circular economy, closer to this book's definition of circular development. It recognises the linkages between looping, ecologically regenerative and adaptive actions. It also territorialises these activities. Motivations for adopting this strategy are economic, environmental and social (solidarity). The social benefits are extremely important in this case. Paris demonstrates three pathways for circular development. The first pathway encourages a city-regional approach to looping construction materials, food and water. It also co-ordinates the strategic reuse of sites (e.g. *Paris Reinvented*). The second pathway adopts a grass-root, temporary-experimental approach to circular development (e.g. Les Grand Voisins, Bellastock), through the adaptive reuse of sites. The third pathway demonstrated by Clichy Batignolles uses a planned eco-district to demonstrate and test the application of the three circular actions in a new build development. The greatest challenge to circular development in Paris appears to be the cost of land and lack of available land. Nevertheless, circular development benefits from considerable government support in Paris, through regulatory and funding mechanisms. The spatial plan, public procurement and government funding (e.g. eco-districts, "Paris Culteurs", "Frigos Solidaire") are successfully leveraging circular development.

Notes

1 For example, the Cinq Toits Project (a former police residence, repurposed as a shelter for the homeless and refugees) and the Parmentier (an electricity substation now used by an art collective).
2 For example, the Archipel, a former nineteenth-century convent, which was converted into an emergency accommodation centre in 2012 for vulnerable women and children.
3 Morland was previously an administrative building. The architect David Chipperfield proposed a vibrant, multi-use building which connected to the river. The multiple-uses ensured that space was occupied 24-7. Thus, the space was fully utilised but also flexible and adaptable. The rooftop was urban farm, which helped to create a local food system. It also helped to begin to restore ecosystem services in the local environment.
4 Mille Arbres was built over a ring expressway to connect two disjointed areas of the city, while retaining a long-distance bus station on an adjacent site. Architect Sou Fujimoto and Oxo Architects created a structure connecting both sides of the city with a public park and restaurant street. It integrated green infrastructure into a heavily trafficked urban environment, helping to reduce air pollution and regenerate the local ecosystem.
5 Massena project was planned for a vacant site with an existing, historic train station (protected by the City of Paris). The winning bid proposed to create a "short food cycle" through the provision of an urban farm and café on the site. All three projects demonstrate facets of circular development (site reuse, adaptive design and ecological regeneration).

PART III

Lessons learnt for circular cities and development

7

IMPLEMENTATION

Pathways, levers and dynamics

Circular cities and development pathways

There are considerable differences in the way "circular" is conceptualised across the cities studied. In London the conceptualisation is economic. In Amsterdam a dualism appears between circular economy and circular development, although the linkages between the two are acknowledged. In Paris and Stockholm, a much more holistic conceptualisation of circular development has emerged. Circular actions are most clearly territorialised in Paris and Amsterdam. There is also some variation in motivations for becoming circular cities. The emphasis on social solidarity as a motivation for circularity demarcates Paris's motivations for circular development from the other cities studied.

The case studies also demonstrate several common features. First, there is a tendency to focus on circular economy rather than circular development in circular strategies. Second, the motivations for doing so are largely economic and environmental. Third, there is a propensity to concentrate on looping actions, rather than regenerative and adaptive actions within the circular strategies (although links maybe made). These looping actions include the reuse, recycling of materials and water, or energy and heat recovery. Fourth, circular cities focus on the localised looping of organic (including soil, food, sewage) and construction waste streams, because these offer greater economic opportunity. Fifth, circular cities replace existing grey infrastructure with recyclable, adaptable and blue-green infrastructure to aid the looping of materials and water. Finally, there are also commonalities in the levers for circular transformation. These tend to be capacity building, regulatory and fiscal instruments, leveraged by land or financial incentives. The cities demonstrate there are several circular development pathways, even within a European context. Amongst the case studies three circular development pathways have emerged: city-regional, eco-district and temporary-experimental pathways.

I *City-regional pathway*

A city-regional pathway has been adopted in Amsterdam, Stockholm and Paris. This takes two forms. First, where development facilitates the localised looping of resources (organic, waste-water and construction waste) within the city-region. Closing these resource loops offers significant economic opportunities. It also increases urban resilience (by addressing resource security) and improves the health of the urban ecosystem (cleansing soils and water). It is supported by the spatial development plan.

Second, where strategic programmes have been set-up to encourage the development of all circular activities by repurposing sites and buildings across the city-region (e.g. Paris Reinvented). These programmes enable the city to adapt to the changing urban context and often incorporate ecologically regenerative actions. The key levers for transformation include public land, funding and databases linking projects with appropriate spaces. Both city-regional pathways impact on land-use, urban form and infrastructure.

II *Temporary-experimental pathway*

Grass-root, temporary experiments provide the opportunity to test the feasibility of circular activities for a short-time period in space-scarce cities (e.g. de Ceuvel, Les Voisins, Bellastock, Brixton experiments). This development pathway relies on the use of temporary sites and the engagement of community and small businesses to implement circular activities. It is facilitated by temporary planning permissions and in some cases limited funding. However, a lack of affordable space and competition for sites is a problem, particularly in ensuring a sustained circular transformation.

III *Eco-district pathway*

Planned eco-districts (QEOP, SRSP, Hammarby and Clichy Batignolles) have resource looping, ecologically regenerative and adaptive capacities incorporated into their design, infrastructure systems and development processes. Circular principles have been adopted by QEOP, SRSP and Clichy Batignolles in the construction and disposal processes; the design of infrastructure; the water supply system and the ecological regeneration of the sites. All of these projects are large-scale, new build developments on prestigious sites. The key levers for transformation include significant public funding, release of public land and regulation (planning conditions and contractual agreements).

Thus, the case studies begin to inform a typology for circular development pathways. This typology can be tested and developed through further investigation across the existing and emerging circular cities in Europe.

Levers for circular development

Amongst the case studies a common set of levers for circular development have also emerged (Figure 7.1). Policies and goals provide the strategic vision for the cities to adopt a circular development pathway. These inform the local decision-making process and help to prioritise circular actions. This is particularly important when there are competing priorities and pressures on scarce land. Planning provides the main tool for implementing circular development. Contractual agreements are the legal instrument used to enforce it. These are particularly applicable to publicly owned land. In Amsterdam, circular land release and contracts have been used to encourage the adoption of circular development.

Spatial plans focus on development and bring together the three circular actions. They can be used to allocate land for circular activities (e.g. Amsterdam) or to ensure the mixture of uses, urban form and infrastructure support circular activities (e.g. Stockholm). It is important that the synergies between circular actions are acknowledged in the plan (e.g. Amsterdam) or through linkages identified across several plans (e.g. Paris). Where actions are dealt with separately (e.g. London Plan) the important synergies (and conflicts) which emerge from taking a more holistic view are lost. Planning conditions can also impede circular development. Moving from a prescriptive to a performance-based approach enables a range of responses to achieving circular development goals (exemplified by De Ceuvel).

Temporary planning permissions are also useful tools for enabling low value circular activities, particularly in land scarce cities (e.g. Paris and London). They provide an opportunity to understand issues surrounding implementation and how to best facilitate circular development. The problem with temporary permissions is that the temporality itself reduces the potential for up-scaling. This might be tackled through more systematic provision of space across the city for circular activities.

Flexible and collaborative planning (as used in Stockholm Royal Seaport) encourages the emergence of adaptable spaces and buildings, alongside engaged communities, who take an active role in the provision of the built environment. This combination of adaptable urban form, self-organising communities and developed learning networks help to deliver greater resilience in cities and reduce the wastage of infrastructure and land.

Planning needs the assistance of the other levers to ensure that circular development pathways emerge. Capital and operational subsidies have been used across all four cities to encourage the transformation of business models, infrastructure and service provision. They have also been used to support circular

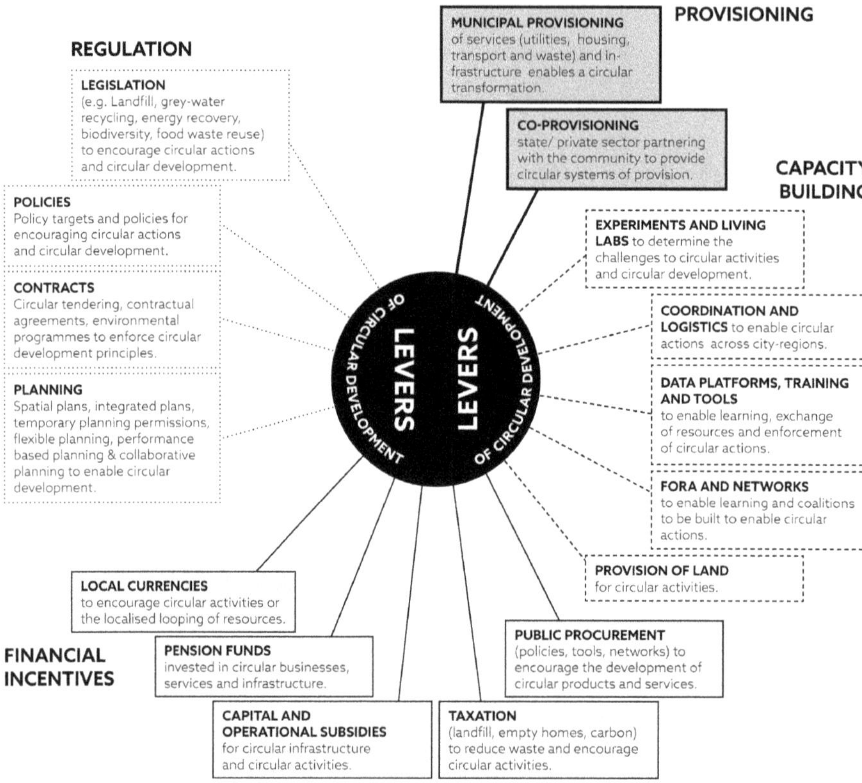

FIGURE 7.1 Levers for circular development.
Source: Authors own produced by Draught Vision Ltd.

social practices (e.g. urban farming, food-reuse). These are the most popular economic instruments for transformation.

Taxation – landfill tax (across Europe), empty properties tax (London) and carbon tax (Stockholm) – has encouraged a shift away from resource wastage and the use of fossil fuels (and greater use of recovered energy and heat). Taxes have produced changes in social practices and systems of provision. Local taxes could be used to support the circular transformation process through investment in infrastructure (as in Stockholm). Public investment may also offer support for circular transformations (e.g. green investment of public pension funds in London). Public procurement is being used in London, Amsterdam and Paris to transform services and infrastructure, to enable circular development. For example, Amsterdam requires that reused baked bricks are used to construct 100% of the public realm works in the city. Paris has developed a tool for calculating the ecological footprint of procurement, which is proving useful for encouraging and monitoring the implementation of circular procurement policies. The use of local currency is perhaps the most innovative economic instrument for encouraging circular activities, tested in Brixton and De Ceuvel.

Capacity building increases awareness, understanding and expertise amongst actors and builds networks (e.g. supply chains) to enable circular development to take place. City councils may provide space in existing buildings for circular businesses (e.g. LWARB's Accelerator Programme) or land for circular development (e.g. Buiksloterham, Amsterdam). They may help to build fora for the exchange of information and networks for delivery (e.g. CE100, Circular Economy Club, C40 cities, Eurocities, Sustainability Cloud); provide training programmes (e.g. circular innovation programme in Amsterdam, London's Accelerator Programme); encourage labelling (e.g. for refurbished and second-hand goods in Paris) or help to establish data platforms (e.g. PUMA, Circle Scan, Sol-dating) and tools (e.g. green index, ecological procurement tool, lifecycle analysis tool) which enable implementation. Urban experiments can be seen across all four cities (e.g. Fablabs, Living Labs). These are used to determine modes of delivery and challenges to implementing circular development pathways.

Municipal provisioning powers could also enable cities to directly control the manner in which services (energy, water, waste, transport) are provided and the type of infrastructure built (as is the case with the water supply in Paris). If a city also prioritises circular development, this can be used to leverage the transformation process. Nowadays, there are a limited number of cities with provisioning powers, because of the shift towards privatisation of services. Stockholm provides a good example of a city, which lost its provisioning power for energy, which has created difficulties in replication of the ecocycles model.

Co-provisioning enables the community to work with public and private actors to deliver circular development. It is exemplified by the community projects in Brixton; the bottom-up development process in De Ceuvel and the food reuse and urban farming schemes in Paris, Amsterdam and London. Co-provisioning engages a range of actors, which splits the burden of cost and increases local support for circular activities. It also helps to encourage the adoption of circular practices at a local level.

System dynamics

The case studies provide evidence that all three circular actions work synergistically together to deliver circular development (Figure 7.2). Looping actions support ecological regeneration and help to build urban adaptive capacity. Looping actions (recycling soil, organic waste, grey and black water) remove pollutants from the ecosystem and enable growth of vegetation (e.g. SRSP and QEOP), resulting in ecological regeneration. Recycling sites and infrastructure (looping) enables the city to adapt to the new demands being placed on it; thus, it becomes more adaptive (e.g. Pop Brixton, Les Grand Voisins). Recycling black- and grey-water and recovering energy from organic waste (looping) can increase urban resilience to resource scarcity (e.g. Paris, Amsterdam). For example, Amsterdam aims to extract proteins from waste-water to feed its population, which

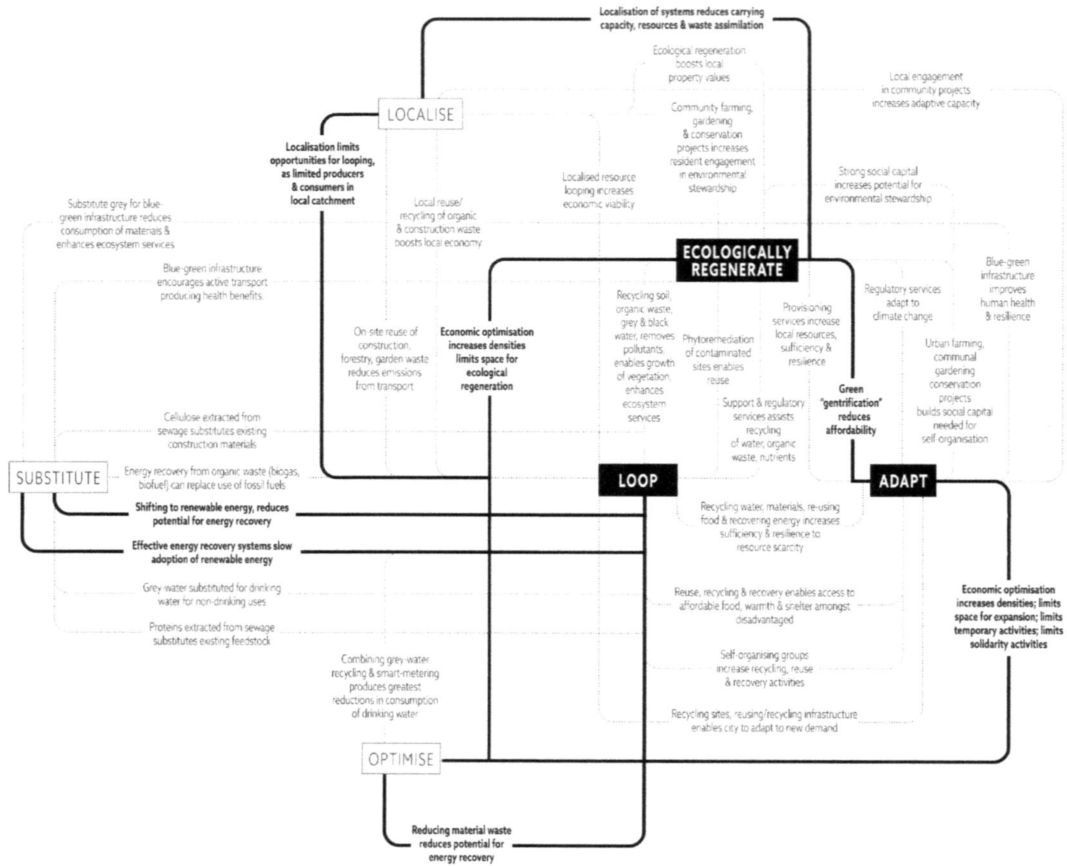

FIGURE 7.2 Dynamics between circular actions and other strategies.
Source: Authors own produced by Draught Vision Ltd.

will increase resilience to food shortages. Heat recovery, reuse of food and properties (looping) enable access to affordable food, warmth and shelter amongst the disadvantaged (e.g. Paris, Brixton), and thus builds urban adaptive capacity.

Ecological regeneration also reinforces looping and adaptive actions. Healthy ecosystems are more effective in recycling of water, organic waste and nutrients (looping). This increases self-sufficiency (clean water, food) and adaptive urban capacity. Human health improves in healthy ecosystems, which increases the population's resilience. Phytoremediation of contaminated sites (ecological regeneration) enables the reuse of land for activities including water storage and urban agriculture (e.g. QEOP, SRSP). This increases urban productive capacity and thus self-sufficiency. Community involvement in gardening, agriculture and conservation projects (as seen in Brixton, Paris, QEOP) helps to build natural, social and human capital within communities, which increases resilience. In addition, all cities studied incorporate blue-green infrastructure to enable adaptation to climate change and improve the health of inhabitants. However, "green gentrification" can produce inequalities, as poorer groups are forced out by high property prices. This creates a conflict between ecological regeneration and a community's ability to adapt to issues such as urban poverty, food scarcity and health problems.

There is also evidence to support synergies between circular actions and other urban strategies (localisation, substitution and optimisation). There is a strong positive relationship between localisation strategies and resource looping, as demonstrated in Amsterdam. Localisation increases the economic feasibility of reusing, recycling and recovering energy from organic and construction waste; reduces emissions from associated transportation and helps to build the local economy. Localisation strategies can also enhance adaptive capacity in communities. Local engagement in community projects, collaborative planning and learning networks increases local capacity to adapt to changing needs (e.g. Brixton, De Ceuvel). The local focus can also help to build community support for environmental stewardship, especially where the benefits of local actions can be seen locally (QEOP, Brixton and De Ceuvel). This builds adaptive capacity, stronger local social and learning networks, which can be mobilised to tackle other issues (e.g. Transitions Town Brixton).

Optimisation strategies and looping can demonstrate a synergistic relationship. For example, smart metering of potable water can encourage greater greywater recycling (as seen in Paris). Substitution strategies demonstrate a synergistic relationship with ecological regeneration. For example, green corridors (ecological regeneration) can encourage the adoption of active modes of transport (substitution strategies) as demonstrated in QEOP. Looping actions produce alternative materials and energy sources which substitute for existing options. For example, biogas in Paris, produced from the anaerobic digestion of organic waste (looping), is substituted for gas in the city heating system (substitution strategies). Cellulose and proteins extracted from sewage in Amsterdam (looping) are used for building materials and feedstock (substitution strategies). Blue-green infrastructure is used to substitute for grey infrastructure (as is the case for Paris's water system), which reduces material consumption and enhances ecosystem services (ecological regeneration).

Of course, there are also conflicts between circular actions and other urban strategies. For example, Stockholm demonstrates how substitution strategies may clash with looping actions. The city has been unable to shift towards renewable energy and recycling systems because of their potential to undermine waste-to-energy systems. Localisation may limit the opportunities for resource looping, if there are limited producers and consumers of waste products within a locality (as highlighted in Brixton). It may also limit the carrying capacity of a system, its ability to produce the resources it needs and assimilate the waste it produces. This is why Amsterdam and Paris are both taking a city-regional approach to construction and organic waste.

Optimising strategies reduce material waste, which can undermine looping strategies (e.g. ecocycles in Stockholm). All four cities demonstrate how optimisation strategies may also clash with circular actions, particularly when applied to land. The optimal use of land (at least from a market perspective) maximises value, by building luxury residential and commercial development on empty sites. This prevents low-value circular activities in cities. Thus, there is limited land available for the industries needed to create industrial symbiosis; for green space and urban agriculture; for solidarity projects and providing space to allow infrastructure to adapt for new needs. The relationships between sharing and circular strategies have not been determined through the case studies. Thus, more research into these synergies and conflicts is needed to determine which development strategies are likely to operate together successfully in circular cities.

Summary

The case studies have begun to demonstrate how circular development manifests in practice. They highlight the differences between circular economy and circular development, and have begun to produce a typology of circular development pathways. The case studies have also demonstrated a range of levers which can be used to encourage circular development. They have investigated how circular actions taken in the city interact and how they operate alongside other urban strategies. Much more needs to be learned, but these findings provide a good grounding for understanding how circular development might be successfully implemented in practice.

8

REASONS FOR ADOPTING A CIRCULAR DEVELOPMENT PATHWAY

The transformation into a circular city is likely to be costly and disruptive. It will require a wholesale shift in the way we plan, design and manage our cities. It will also necessitate changes in social practices, lifestyles and systems of provision. The support for this transformation, from politicians and those inhabiting the city, will be essential. This chapter presents the evidence for adopting a circular development pathway gleaned from the case studies and from the wider research.

Reduce resource consumption and wastage

All three circular actions can help to reduce resource consumption. Reuse and recycling of construction materials and adaptation of infrastructure reduces waste, as well as material and energy consumption. In Amsterdam, it is estimated that the high-value reuse and recycling of construction waste would save 500,000 tonnes of materials per annum. Recycling soil also reduces wastage and the need to import topsoil. Soil-washing resulted in 90% of excavated soil in the Queen Elizabeth Olympic Park (QEOP) being reused on site, which reduced transportation and associated emissions. Integrated closed-loop systems can reduce resource consumption and waste. In Hammarby, Ecocycles reduced fossil fuel consumption by 28–42%, water consumption by 41–46% and waste going to landfill by 90%. These are very significant savings.

Recycling proteins from waste-water to produce animal feed-stock, as trialled in Amsterdam, reduces waste. It also reduces the land area and water needed to grow crops, as well as the energy costs associated with production and transportation. Heat recovered from buildings (De Ceuvel) or from waste-water treatment (Hammarby) reduces energy consumption. However, sometimes looping can increase resource consumption, due to the rebound effect. For example, the renewal of the grey-water recycling system in Paris might actually increase overall water consumption, unless it is metered.

The ecological regeneration of vacant brownfield sites can limit the consumption of land outside the city for new development. Vacant properties and sites are being identified in Paris, Amsterdam and London to reduce under-utilisation. The adaptation of existing infrastructure for new uses reduces resource wastage, and the need to build new or demolish existing infrastructure (e.g. Paris Reinvented). Temporary planning permissions are used in all three cities to encourage the reuse of space and infrastructure. Flexible and multi-use spaces incorporated into buildings and urban form can reduce land consumed by urban activities. Flexible buildings in Amsterdam and flexible planning in Stockholm are used to this end.

Ecosystem services

Regenerative and looping actions improve ecosystem services.[1] For example, off-grid rainwater harvesting and grey-water reuse systems have been adopted in Queen Elizabeth Olympic Park (QEOP) and De Ceuvel (DC). These provide a clean water supply and reduce problems with drought and flooding on site. Grey-water recycling can also enhance river ecology, by preventing sewage overflows as in Stockholm Royal Seaport (SRSP). The provision of green and blue infrastructure regulates the urban climate (Corvalán et al., 2005; Peng et al., 2012). This in turn reduces energy consumed by heating and cooling systems.

Green corridors, like those adopted in SRSP and QEOP, allow for the dispersal of plant and animal species. This increases biodiversity. Bioremediation (with bacteria) and phytoremediation (with plants) have been used to decontaminate soil and water in SRSP, QEOP and DC. This increases land availability in urban locations, potentially protecting the hinterland from development. Green infrastructure can also help to regulate noise pollution and air pollution (Chaparro and Terradas, 2009). It has been used to this end in all four cities. It can also produce raw materials for use in the bio-economy. For example, water-weed is harvested and used to create construction materials in Amsterdam. Blue-green infrastructure also provides important cultural services,[2] preserving cultural identity and a sense of place.

Reducing greenhouse gas emissions

Circular actions can reduce greenhouse gas emissions. Energy recovery from organic waste helps reduce greenhouse gasses produced from landfill and decarbonises the energy supply. For example, biomass is used to produce biogas in Paris, which is injected into the heating system to decarbonise it. It also reduces waste going to landfill. Ecocycles (energy recovery system) in Hammarby has reduced CO_2 emissions by 29–30%.

The Dutch government recognises the contribution an improvement in efficiency in raw material and material value chains could make to cutting CO_2. They estimate a 9% reduction in emissions (Blok et al., 2017). The case of Amsterdam demonstrates that looping construction and organic waste flows can make a significant contribution. The circle scan project calculated that high-value recycling and reuse of construction waste could save 75,000 tonnes of CO_2 per annum, whilst dismantling and separation of components and materials from buildings could save 100,000 tonnes of CO_2 per annum (Bastein et al., 2016).

Looping organic waste could also make significant savings. The circle scan project calculated that cascading of organic waste flows (300,000 tonnes of CO_2 per annum), organic waste separation (100,000 tonnes of CO_2 per annum), recovering nutrients from organic waste (100,000 tonnes of CO_2 per annum) and the establishment of a bio-refinery hub (300,000 tonnes of CO_2 per annum) could save 800,000 tonnes of CO_2 per annum (ibid).

Green infrastructure can be used to sequester CO_2 (McPherson, 1998; Nowak, 1994) and fix pollutants including O_3, SO_2, NO_2, CO, and PM10 (Chaparro and Terradas, 2009). All four cities encourage the inclusion of green infrastructure into the urban fabric as a strategy for tackling CO_2 emissions. The adaptation of infrastructure within city-regions can also reduce energy consumed in the construction, transportation and disposal processes, thus reducing CO_2 emissions. It is estimated that smart adaptive building in Amsterdam could save 300,000 tonnes CO_2 per annum.

Resource sufficiency

Urban resource sufficiency is also affected by the three circular actions. Localised looping actions reduce resource wastage and increase the supply of resources within the city. For example, in Paris efforts are being made to recycle soil waste locally. The sol-dating platform recovers more than 30,000 m^3 of soil in the Île-de-France region, creating an alternative to the distant supply of topsoil. Thus, the city can become soil-secure. Nutrient recovery from residual food in Amsterdam, for reuse (by restaurants or foodbanks) or composting, will capture 95% of the nutrients lost currently. The power-to-protein project extracts ammonia from sewage, and could reduce Amsterdammer's reliance on external protein sources. These strategies can help to tackle food security.

The recycling of grey-water and black-water, as demonstrated in De Ceuvel and Queen Elizabeth Olympic Park, can also reduce the consumption of potable water. This can help to tackle water security in water scarce environments. The ecocycles system in Stockholm reduces the city's reliance on external energy sources, thus increasing energy security. Regenerating local ecosystems also enables the local production of raw materials, food and energy. This reduces the need to import resources and is being encouraged in all four cities. For example, local urban agriculture is supported by *Incredible Edible* in London and *Paris Culteurs* in Paris. Sufficiency helps all four cities to become more resilient to resource insecurities. Also, introducing urban agriculture and energy generation into the city reduces the land-take for food and energy uses beyond the city-limits.

Adapt to environmental change

Adaptive capacity may be built within communities, enabling inhabitants to self-organise to tackle both climate adaptation and mitigation. This was the case in Transition Town Brixton. The movement bought people together in the area to create their own solutions.[3] This was coordinated by Transitions Town and mobilised through the local currency and allocation of space. These community actions strengthened and expanded social and learning networks, increasing the capacity of the community to self-organise and act. The variety of projects and the local currency were key to engaging a wider section of the community.

Urban form and infrastructure also enable the cities to adapt to climate change. The inclusion of blue-green infrastructure in all four cities was highlighted as being instrumental in climate change adaptation (e.g. urban cooling, flood management). In Stockholm and Amsterdam new developments have also been designed to adapt to sea-level rise. Increasing resource security in Amsterdam (food, construction materials) and Paris (food, energy, soil and water) will also help them to adapt to climate change.

Increase environmental awareness

Regenerative and looping actions at a local level can help to renew society's connection with the natural world as well as understanding of resource cycles and product life-cycles. Repair cafes (in Paris, Amsterdam, London), labelling of second-hand or refurbished goods (in Paris), food reuse cookery classes (e.g. Brixton Pound Café) all raise environmental awareness, change social practices and encourage looping actions.

The provision of green and blue infrastructure in cities can also help to reconnect people with their local environment (Collado et al., 2013; Ward Thompson et al., 2008). Indeed, community engagement in the restoration of green spaces and waterways (e.g. park conservation in QEOP, Paris Culteurs) raises environmental awareness and changes social practices. It increases environmental stewardship, environmental awareness and the value the public places on local resources. Green and blue infrastructure and the localised looping of water and materials have helped to demonstrate the importance of the human–nature relationship. The degradation of the urban ecosystem resulting from human behaviour is better understood when experienced locally. Off-grid solutions to water treatment and energy production (as used in De Ceuvel) provides the public with a better understanding of the impact of their consumption patterns.

The public may also engage in environmental behaviour for social reasons (e.g. solidarity, opportunities to socialise). People may become involved in food reuse projects (e.g. Brixton Pound Café, Freegan Pony and community fridges in Paris) for solidarity reasons. Their engagement is socially motivated, but it also reduces food waste. The public also engage in projects because of opportunities to socialise, for example, urban farming and repair cafes in Paris, London and Amsterdam. Yet engaging in these activities has environmental benefits. In both instances, an increase in environmental awareness is a by-product of involvement.

Social and economic benefits can also provide strong motivation for circular transformation and are particularly attractive to politicians and to the public. Social benefits can be subdivided into health and community benefits. Both are fundamental to the well-being of urban populations.

Health benefits

It seems that health benefits solely relate to *regenerative* actions. Put simply green and blue infrastructure can improve air quality, reduce heat stress and noise pollution and create spaces for recreation and relaxation, all of which have significant health benefits. Vegetation can reduce exposure to pollutants (PM2.5, PM10, O_3, NOX, SO_2, VOC and toxic metals) by directly removing or dispersing them. Barcelona's trees and shrubs have removed 305.6t yr^{-1} of pollution from the air, saving health costs valued at €1,115,908 a year (Chaparro and Terradas, 2009). A study in Carlisle showed that trees could reduce childhood asthma by up to 29% (Carlisle City Council, 2011).

Green and blue infrastructure can also be used to tackle heat stress (Corvalan et al., 2006; Peng et al., 2012; Zoulia et al., 2009). Research has shown that during heat waves air quality is degraded and the concentration of pollutants increases (Corburn, 2009; Harlan and Ruddell, 2011). It is estimated that up to 12% of air pollution problems in cities are attributable to the heat island effect (Forest Research, 2010). The most vulnerable populations are the elderly and young children. Individuals with pre-existing health conditions, such as cardiovascular and respiratory diseases, are at greatest risk (Hallegatte et al., 2011).

Heat is also an occupational hazard, especially for outside workers (Kovats and Akthar, 2006). Shade from trees and short vegetation lowers temperatures (Bowler et al., 2010). Increasing the canopy cover may reduce air temperature by 1–3 degrees. Green roofs may also decrease heating and storm run-off (O'Neill 2009). Density and type of vegetation play a role in reducing temperature during heat waves (Meier and Scherer, 2012; Skelhorn et al., 2014). Deciduous trees have a greater cooling capacity than coniferous trees. It is estimated a 10% increase in green areas across cities could mitigate an expected temperature rise of 4°C (Gill et al., 2007).

Exposure to excessive noise is considered the second-worst environmental cause of ill health after PM2.5 pollution (WHO, 2011). It produces various negative impacts, ranging from the minor discomfort produced by sleep disturbance to serious cardiovascular problems. Noise is caused by transportation and industrial activity. Vegetation in urban environments can reduce noise by absorbing or diffracting it (Van Renterghem et al., 2015). There is also evidence that the presence of vegetation influences noise perception. Parks have been used to ameliorate the noise impacts of roads (e.g. Parc des Hautes Bruyères, Paris) and airports (e.g. Buitenschot park near Schiphol airport, Amsterdam).

There is also a relationship between green space, self-perceived health and doctor-assessed diseases (Maas et al., 2009). Local green spaces (within 1 km of people's homes) have a significant effect on mental and physical health conditions. This particularly affects children and people from lower socio-economic groups, who spend most time around their house. Studies in Denmark showed that those living more than 1 km away from green space reported poorer health and health-related quality of life. They experienced more stress than people living closer to a green space (Stigsdotter et al., 2010). Pregnant women living more than 300 meters away from green spaces had higher blood pressure compared to those living closer (Grazulevicience et al., 2014).

Proximity to green spaces also reduced behavioural problems in children, with hyperactivity, emotional and peer relationships problems (Balseviciene et al., 2014). It improved working memory in children (Dadvand et al., 2015). Children with attention deficits concentrated better after a walk in the park. Green school playgrounds improved well-being and diminished physiological stress (Kelz et al., 2013), improved attention, reduced behaviour problems and enhanced factors associated with resilience in children of all ages (Chawla et al., 2014).

The provision of green infrastructure also encourages active lifestyles amongst the wider population (Janssen and Rosu, 2015). Active lifestyles reduce obesity and levels of stress and improve the mental health of those living in cities. Urban farming (at a variety of scales) can also help to tackle obesity. Opportunities to grow fresh food can help improve nutrition. Thus, regenerative actions increase urban inhabitants' longevity, by improving their physical and mental health.

Community benefits

Circular development helps to build local symbiotic capital. Local circular activities strengthen social and learning networks, encouraging reciprocity and mutual aid within communities. This was demonstrated by the food growing and food waste reuse networks in London, Paris and Amsterdam. Circular activities build local expertise and skills (e.g. in Brixton repairing goods, food reuse, generating and managing community energy systems). They may also provide much needed physical infrastructure (e.g. the temporary reuse of empty buildings in Paris). Guerrilla gardening (London and Paris), nomadic animal herding (Paris) and waterway conservation projects (QEOP) can protect ecosystem services and natural resources. Finally, circular activities can generate local financial capital (e.g. urban agriculture, community energy, pop-up businesses in Brixton).

Circular development can help to stabilise and build communities. However, it can also result in gentrification, which prices out low-income groups (Dooling, 2009). This was seen in SRSP and QEOP. Gentrification could be partially addressed through the planning process by ensuring affordable housing and opportunities for low-income groups are included in the development (Rigolon and Németh, 2018). As local symbiotic capital grows, communities become more empowered. This can be encouraged through the allocation of space or funding for community projects. It can be sustained by creating local demand for the resources produced, possibly through the use of local currencies.

Circular activities also often fulfil essential human needs, for example, food reuse schemes (*Freegan Pony* café, cookery classes in Brixton, community fridges in Paris, *Food-Cycle* markets in Amsterdam), homeless hostels in vacant buildings (e.g. *Les Grand Voisins)* and renewable energy projects for social housing (e.g. *Brixton Energy*). All provide opportunities for social interaction alongside enabling access to food, shelter and warmth for the urban poor. In Paris, this is recognised and supported through solidarity policies. Community projects encourage the emergence of social practices which underpin circular development. The motivations for doing so are often social (rather than environmental), but the outcomes are the same. The projects also build the networks for learning and self-organisation critical for creating adaptive communities. This can be seen in Brixton.

Economic growth and jobs

It is widely acknowledged that circular economy can produce economic benefits. Transition to the circular economy in Europe could create up to 1.2 to 3 million jobs and reduce unemployment by around 250,000 to 520,000 (WRAP, 2015). It is estimated the circular economy could be worth as much as £9–29bn for the UK (Eunomia Research Consulting, 2016) and €7.3 billion for the Netherlands annually (Bastein et al., 2013). It could also create 10,000–175,000 jobs in the UK (Voulvoulis, 2015) and 54,000 jobs in the Netherlands (Bastein et al., 2013) and 300,000 jobs in France (Ministry for an Ecological and Solidary Transition and Ministry of Economy and Finance, 2018). The economic benefits provide an important motivation for nations and cities to adopt looping actions. Some attempts have been made to calculate the economic impact of looping actions within specific sectors (construction and biomass and food waste sectors) in cities. For example, in London, it is estimated that circular economy could create 12,000 new jobs by 2030, of which 5% would be in the construction industry (London Sustainable Development Commission, 2015). Tackling the construction waste stream in London could generate economic growth of between £3bn and £5bn annually by 2036 (LWARB, 2015). The capital's circular food economy could add £2–4 billion annually to GDP by 2036 (ibid). It is also expected that reuse, remanufacturing and materials innovation could add at least £7bn to the London economy annually (ibid). In Amsterdam, the circular construction sector has been predicted to produce €85,000,000 annually and 700 jobs. The organic waste (including food waste) sector is expected to generate €140,000,000 annually and create 1,250 jobs (ibid). These figures suggest the economic value and jobs created by looping actions could provide ample motivation for adopting a circular development pathway.

Similar studies assessing the economic impact of ecological regeneration or adaptation are not available. However, we know that the employment opportunities created by ecological regeneration are wide-ranging and reflect the ecosystem services provided (ten Brink et al., 2017). Thus, employment opportunities may emerge in conservation, urban forestry and agriculture, gardens and park management, water management, carbon sequestration, recreation, health and tourism, education, research and development. However, it is very difficult to calculate the economic value generated by the management of ecosystems services (Gómez-Baggethun and Barton, 2013). This is because the economic benefit of ecosystem services is often calculated in terms of the costs avoided (i.e. health, insurance costs) rather than the economic value of businesses engaged in ecological regeneration.

Adaptive actions are similarly broad in terms of the employment opportunities created, connected with urban resilience,[4] the pop-up economy, adaptive infrastructure[5] and co-provision. Thus, it is difficult to enumerate the worth of these actions or the number of jobs created. What is clear is that circular actions will create new economic opportunities across primary (forestry and farming), secondary (manufacturing), tertiary (service) and quaternary (R&D) sectors. This produces new income streams and a diversity of employment opportunities, requiring equally diverse skills. So moving towards circular development could create a more inclusive and stable economy. The broader industrial base could help to increase long-term economic security. More research is needed to quantify these benefits.

Economic efficiency

Looping and adaptive actions can remove redundancies in the urban system by eliminating waste. For example, in designing flexible buildings we can avoid the resource costs (materials and energy) of demolition and construction. Circle Scan calculated that smart (adaptive) design approaches could generate 100 jobs and create a value of €12,000,000 per annum for Amsterdam (Bastein et al., 2013). Equally, recycling (the materials or components) or reusing infrastructure avoids resource wastage and associated costs. Circle Scan calculated recycling and reuse of construction materials could generate 200 jobs and create a value of €23,000,000 per annum (ibid).

Increasing efficiencies in the supply and production processes by reusing or recycling "waste" resources also reduces financial cost. For example, the *soldating platform* in Paris halved inert soil management costs. In Amsterdam, it is estimated that food reuse could create an economic value of €30,000,000 and 150 jobs per annum (ibid). Grey-water reuse can also offer indirect benefits to public infrastructure costs in the form of reduced sewerage flows, reduced treatment plant size, shorter distribution systems and reduced potable water demand.

TABLE 8.1 Economic accounts for the ecosystem services air purification, urban cooling and climate regulation

Ecosystem service	*City*	*Economic value estimates*	*Valuation model*	*References*
Air purification	Barcelona	€1,115,908	Avoided costs (UFORE)	Chaparro and Terradas (2009)
	Chicago	$9.2 m	Avoided costs (CBAT)	McPherson et al. (1997)
	Modesto	$1.48 m ($16/tree)	Willingness to pay	McPherson et al. (1997)
	Sacramento	$28.7 m	Avoided costs	Scott et al. (1998)
	Philadelphia	$3.9 m/year	Avoided costs	Nowak et al. (2007)
Urban cooling/ heating	Chicago	$15/tree (cooling)	Avoided costs (CBAT)	McPherson et al. (1997)
	Chicago	$10/tree (heating)	Avoided costs (CBAT)	McPherson et al. (1997)
Climate regulation	Modesto	$5/tree	Avoided costs (CBAT)	McPherson et al. (1999)
	Philadelphia	$ 9.8 m	Avoided costs (UFORE)	Nowak et al. (2007)

Source: Adapted from Gómez-Baggethun and Barton (2013).

It can help prolong the need for additional potable water sources (Radcliff, 2003). A study found that Brisbane City Council could realise waste-water treatment savings up to $42,000,000 per year if grey-water reuse systems were introduced into urban areas (Jeppesen and Solley, 1994).

Ecological regeneration can help to avoid costs (Table 8.1). It can reduce the economic costs which arise from health problems related to loss of ecosystem services. For example, it is possible to avoid the health costs of treating cardio-vascular, respiratory and mental health problems created by air pollution and noise (ibid) by introducing green walls into urban environments. Vegetation regulates urban heating and thus avoids additional energy costs in the summer season. Land-use change in urban water-catchments, particularly the loss of trees, can lead to the construction of costly water purification plants and flood alleviation measures (Daily and Ellison, 2012). In 2018, flooding caused $82 billion of economic damage globally. Thus, loss of ecosystem services in urban areas has a significant economic cost (De Groot et al., 2010).

Enhancing the value of real estate

Circular development contributes to the improvement of local environments, resulting in economic revitalisation, activation of vacant spaces and increase in land and property values. The ecological regeneration of neighbourhoods through the provision of blue-green infrastructure often increases land and property values

(De Groot et al., 2013; Okvat and Zautra, 2011; Roy et al., 2012). A rigorous analysis of hundreds of New York City gardens demonstrated that opening a community garden has a statistically significant positive impact on the sales prices of properties within 1,000 feet of the garden (Voicu and Been, 2008). This is an impact which increases over time and is greatest in disadvantaged neighbourhoods (ibid). Increase in land and property value also boosts the yield from local tax revenue. Greening neighbourhoods produces the upgrade effect. Thus, surrounding neighbourhoods also introduce green infrastructure into the living environment, setting off a cycle of improvement, which spreads to all sectors of the community (Hall, 2011).

Adaptive actions, particularly the temporary reuse of space, can result in economic revitalisation and boost real estate value (Madanipour, 2018). At the very least temporary activities are a productive use of empty space. Temporary spaces also provide access to low-value uses in competitive economic environments (ibid). These activities promote a perception of vibrancy, which can quickly create interest in abandoned or stalled development sites. This increases their visibility and agency within a neighbourhood (Németh and Langhorst, 2014). This process extracts latent value from temporarily disused sites (Bishop and Williams, 2012). It becomes a valuable urban model, which reduces economic risk, unlocks potential of sites and generates a capital flow, which "does not come into conflict with the immobility of real estate" (ibid). It is argued that temporary projects can lead the way in promoting the "innovation, fluidity and flexibility" needed in twenty-first-century cities (ibid). However, the gentrification process associated with increasing value can also result in social exclusion (e.g. QEOP and SRSP).

Local economic benefits

Circular actions can also produce localised economic benefits. The low value of residual resources requires resource loops are closed locally (as demonstrated by Circle Scan, Amsterdam). Thus, residual resources are reused, recycled, composted or energy is recovered locally, to maximise economic return and ensure economic viability. In Paris, organic waste is converted into compost and used to upgrade soil which is reused locally or to produce biogas which is injected into the local energy system. Capturing value locally helps to grow and diversify the local economy.

Ecological regeneration results in improvement in the local living environment which has a range of benefits (health, well-being, aesthetic attractiveness). This increases the value of local real estate and tax revenue collected, which can be spent on improving local infrastructure and services. Thus, ecological regeneration can produce a virtuous loop. However, the increasing value of real-estate can also create social exclusion. We have observed this in QEOP (London) and SRSP (Stockholm) sites where local people are relocated or priced out of the housing market. Rising real-estate values will also reduce the opportunities for other low-value circular activities, except on temporary sites.

Ecological regeneration enables the production of resources locally – clean water, soil, food – which may expand and diversify the urban economy. This is exemplified by Paris, which aims to increase local food production in the city-region. It can also increase inclusivity, offering employment opportunities to people with a range of skills and economic means. However, in some instances land prices exclude community enterprises. This is an issue in Paris, where community agriculture occurs on small temporary sites, over-shadowed by the large-scale commercial alternatives.

Summary

This chapter begins to provide an evidence base for adopting circular development in cities (Figure 8.1). Circular development produces environmental, community, health and economic benefits. Ecological regeneration produces the greatest number and most diverse set of benefits. These are complemented and amplified by the benefits accrued from looping and adaptive actions. Attempts have been made to measure and quantify some of the benefits; however, the data is currently very limited. The benefits emerging from taking circular actions have been investigated separately. Yet systemic thinking suggests these actions will interact and new synergistic benefits (and conflicts) may emerge. More research quantifying the benefits, identifying the synergies and conflicts arising from adopting circular development is needed.

There is potential for the inequitable sharing of the benefits, derived from circular actions, across society. For example, access to green space is often restricted amongst poorer urban groups, who live in high-rise blocks or higher density social housing developments. Yet the health benefits of locally accessible green space to these groups are greatest, because they tend to remain in their local area. This inequality is compounded by the fact that the maintenance of public green spaces is increasingly becoming a problem, with limited public funding. Thus, public green spaces are lost or poorly maintained, which further limits access to the urban poor.

"Green gentrification" increases land and property values, which excludes lower-income groups. The inclusion of green technologies (which generate renewable energy, reuse grey-water and heat or increase energy efficiency) and green space into new developments reduces affordability. Yet the socially excluded could benefit most from green space and affordable utilities. The case studies also show that the "solidarity" circular activities (e.g. food reuse cafes, community urban farms and the provision of temporary accommodation for the homeless) which particularly benefit the urban poor are more likely to fail, because they are not commercially viable.

It is predicted that circular economy will provide a variety of new job opportunities for a range of skills sets. However, it is more likely that lower-income and poorly educated groups will be employed in activities which could be detrimental to their health (e.g. recycling e-waste). Without educational and training programmes these groups will continue to be excluded from the highly skilled

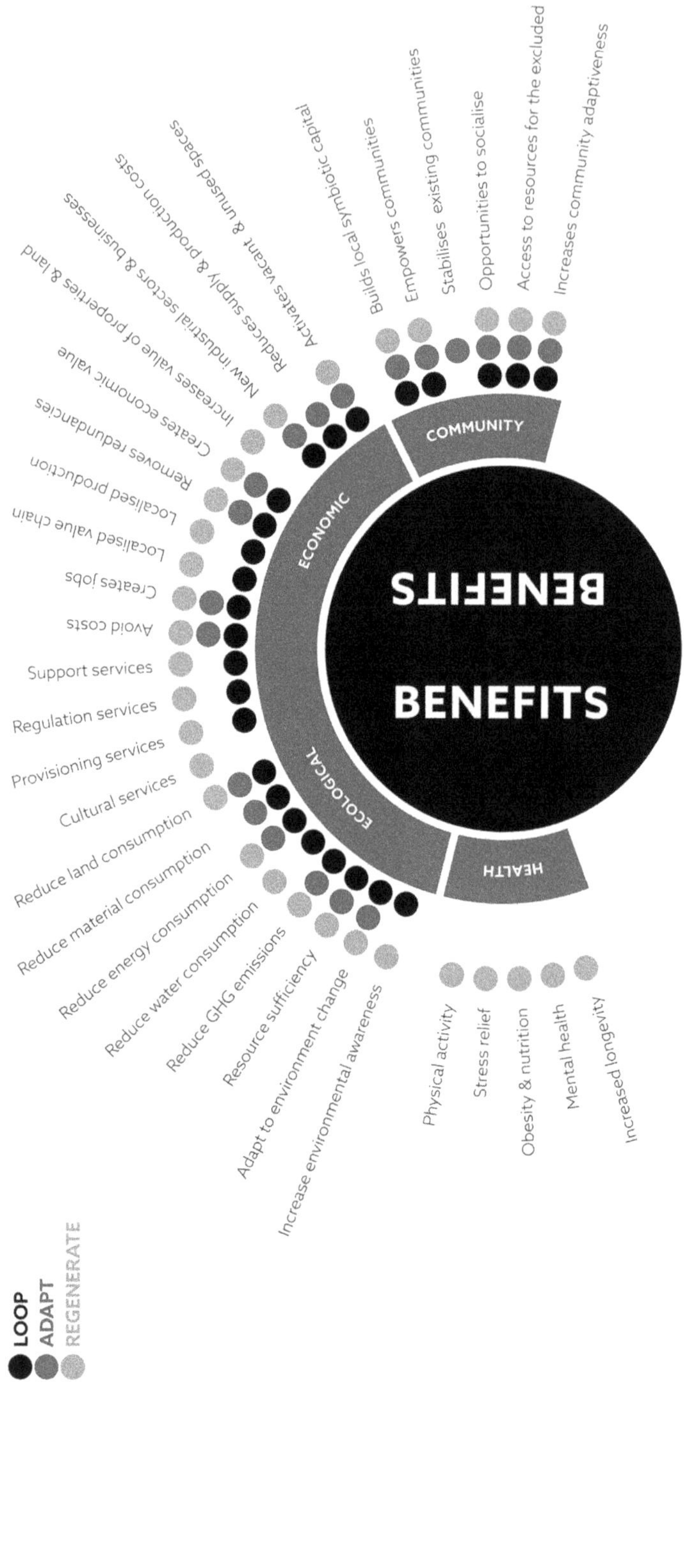

FIGURE 8.1 Benefits of circular development.
Source: Authors own produced by Draught Vision Ltd.

circular jobs. Finally, the adaptive capacity amongst low-income groups is likely to be less. This is because they have neither the financial nor human capital to self-organise and adapt to changes in their environment. This makes low-income groups more vulnerable to change, which could have detrimental impacts on them. The potential social inequalities resulting from taking a circular approach to development require further investigation.

However, it could be argued that these problems result from political decisions made relating to circular development pathways. For example, the recycling of e-waste can be regulated to remove harmful health impacts. Government can provide public funding to support the maintenance of public green spaces, the inclusion of green technology in social housing and solidarity projects. Land could be designated for green space and other circular activities within low-income communities. All these problems could be tackled given the political will, through greater public funding, allocation of public land and regulation (as demonstrated in Paris). Nevertheless, the more comprehensive monitoring of benefits and disbenefits of adopting circular development, across social groups, can help to enable more informed policy choices.

Notes

1 Ecosystem services include support, regulating, provisioning and cultural services.
2 Cultural ecosystem services include spiritual or religious enrichment, cultural heritage, recreation and tourism, aesthetic experience.
3 Projects include generating renewable energy, growing food, recycling goods and reusing food waste.
4 Monitoring and early warning systems; disaster planning; risk assessment and insurance; communications and the building of action networks.
5 Engineering and design, construction, management and maintenance.

9

CHALLENGES TO IMPLEMENTING CIRCULAR DEVELOPMENT

Currently, there are pioneering cities across Europe, experimenting with circular development pathways. However, in order for these experiments to sustain and scale-up, some fundamental changes in the development regime will be required. In this chapter we explore the challenges to transformation.

Transforming the economic system

The economic system is a major challenge to the implementation of circular urban development. Resources and ecosystem services are under-valued by the market. The true environmental and social cost of consuming resources (i.e. degradation of ecosystem services; degradation of distant communities; loss of irreplaceable, finite resources) is not paid by the market. This is due to the inherent flaw in neoclassical economic model, which supposes that resources are infinite. Ecosystem services are viewed as a common asset which are non-excludable, so these too are under-valued and often over exploited. A seismic shift towards an economic system which internalises the environmental and social costs of resource consumption and societal benefits offered by circular activities is key to underpinning the success of circular development pathway and a fully functioning circular economy.

Recycled, reused and reprocessed materials; decontaminating brownfield sites; recycling grey-water, maintaining blue and green infrastructure; and constructing adaptable infrastructure all come at an additional cost. Virgin materials, potable water and greenfield sites are comparatively cheap. Blue and green infrastructure, although cheaper than grey infrastructure in terms of capital cost, can be more expensive to manage and maintain long-term. Adaptable infrastructural systems, adopting different technologies and designs, will in the short-term incur additional capital costs. In the long-term it is likely to be cheaper to design adaptably because it avoids the costs of demolition and disposal. However, consumers are not prepared to pay these additional costs.

The challenge is further amplified by consumers' concern about the quality of reused products and materials; and the operational and commercial performance of repurposed, refurbished and adaptable infrastructure (Bastein et al., 2013; Bullen and Love, 2010; Wilcox et al., 2016). For example, the parity in price for potable and non-potable water in Queen Elizabeth Olympic Park (QEOP) rendered the black to grey-water recycling system financially unviable. Yet the societal costs of drought or sewage contamination of potable water supply during flash-flooding, which could be avoided by adopting this system, are not factored into the decision-making process. Similarly, developers tend to avoid brownfield sites in cities because of the potential cost of decontamination and the difficulties of consolidating sites. The loss of ecosystem services resulting from development on greenfield sites or the cost to society of building in risky areas (e.g. on floodplains) is never factored into the equation.

Energy recovery from waste appears to be the exception. It is an economically viable alternative to disposal and recycling, as there is a sustained demand for energy. Also in Europe escalating landfill costs and scarcity of landfill sites make disposal an expensive option. The future uncertainty created by resource price volatility (particularly for material recyclates) and changes to global supply chains makes investment in recycling systems riskier (Swickard, 2008; Velis, 2015). So energy recovery systems are seen as a less risky financial investment.

Circular actions produce long-term, societal benefits (under-valued by the market) and require long-term investment. It is often difficult to finance circular projects, because investors and infrastructure providers are looking for short-term returns, driven by short investment cycles (Byström, 2018). This is a major challenge to overcome. For example, building capacity into systems for adaptation and ecological regeneration, through the allocation of space in cities for this purpose, will result in the short-term, economic under-utilisation of land. This will produce low economic yields for investors in the short-term. However, in the longer-term it will provide valuable ecosystem services which have societal benefits. It will also provide room in cities for the expansion or modification of existing infrastructure, which is more cost-effective than demolition, disposal and construction.

Integrating whole life-costing and valuation of ecosystem services into business models could help to overcome this. However, the split-incentive along the value chain tends to undermine this approach. Those making the up-front, capital investment do not profit from the long-term societal benefits. This might be addressed by adopting service-based approaches, which require infrastructure providers to become service providers. For example, house-builders would become responsible for maintaining accommodation long-term, and would benefit from the flexibility spaces for expansion could offer.

Sunk costs in infrastructural systems also prevent replacement. However, in new developments or when decision makers are considering choices for replacement, adopting circular alternatives maybe more financially viable. For example, in Paris the government is considering replacing the existing greywater infrastructure (grey infrastructure) with blue-green infrastructural solutions. They are already testing these options in parts of the city, including Clichy Batignolles. Nevertheless, the sunk costs are a real challenge in many cities, where infrastructure is not nearing the end of its lifetime. The cost of replacement is a very significant barrier to overcome.

Land value and competition for space in many European cities create a considerable challenge to implementing circular development. Circular land-use is under-valued. The societal benefits offered by these activities are not factored into their value. Thus, it is impossible for circular activities to compete with high-value activities, such as luxury residential and commercial development. We have seen this in all the case studies. Circular activities are often only granted temporary permissions. This temporality is unlikely to encourage scaling-up and thus to result in new development pathways. Some compromises have been found. For example, the activities in Les Grand Voisins were moved to an adjacent site once the redevelopment started. Another compromise solution was found in Brixton where a high-value, commercial development was built around the urban farm at Loughborough Junction. The challenge is to enable land use which promotes circular activities, to last beyond the temporary permissions, and scale-up across cities.

Globalisation of resource flows and impacts of consumption reduces people's awareness of the effect of their consumption decisions on society and the environment. The localisation of resource flows and impacts through the creation of local economic systems could help to address this challenge. It will reduce the resources consumed by transportation and associated emissions. It can also help to promote circular activities. The low value of recyclates and reused materials means that looping actions are often locally constrained. However, there are considerable challenges faced in the implementation of a local economic system.

The greatest challenge is the lack of urban producers. Too few urban producers create a problem for local sourcing and resource looping. Often local producers (e.g. industries, urban farmers) cannot compete successfully for space. Yet these actors are needed if "waste" resources heat, materials, goods are going to be reused, recycled or recovered locally. For economic reasons loops have to be closed locally. This creates a major barrier to the adoption of circular development. Of course, planning could intervene in land markets to enable circular activities.

Local currency could be used to encourage circular activities. A local currency is being mobilised in Brixton, but it is a major challenge. The municipality plays a key role in ensuring its success, by generating demand (by paying its workforce and allowing businesses to pay rates using the local currency). Technology has been developed which enables easy transactions. Also a variety of services and goods are provided locally, which accept the currency. The currency supports local production, suppliers and circular activities. However, it takes considerable determination on the part of local actors to make the currency work.

Shifting cultural values

There are major cultural challenges to implementing circular development pathways. Culture influences the values of society, which affect social practices and lifestyles. These influence consumption decisions, environmental praxis and adaptive capacity. There are nine national cultural differences, three of which are likely to effect the implementation of circular development (Hofstede, 2001; Schwartz, 2008). These values are individualism, short-termism and humanity.

Individualism drives resource consumption and wastage. The private consumption of resources (e.g. cars, houses, etc.) demarcates status and provides personal freedom. However, it reduces social capital. This makes collective action to address resource depletion or self-organise to adapt to changing environments more difficult. In contrast, societies with collective values view resources as an asset for the commons, and are concerned about resource over-exploitation. There is greater social capital and willingness to be engaged in community actions for resource stewardship. The social capital helps to build societal resilience and enable self-organisation, learning and a collective response to contextual changes. Thus, circular development sits more easily with collectivist values. Yet, the culture of individualism dominates. However, there are communities of interest through which a collective response is possible (e.g. Transitions Town Brixton, Incredible Edible).

Societal orientation to time will also influence capacity to implement circular development pathways. In western societies, short-termism arguably underpins most political and financial decision-making, largely because of short political and investment cycles. This is a significant challenge to overcome when implementing circular development. A long-term, future-orientated society is more likely to be concerned about futurity and intergenerational equity and the preservation of resources and ecosystems for future generations. This orientation aligns with the goals for circular development. For example, in Sweden, there has been political and financial support for the implementation of closed-loop systems (*ECOCYCLES*) for 20 years. Circularity has underpinned the cultural values of its institutions during this period, which has manifested in policy and practice.

Societies which are highly humane believe in the importance of "others" (people and nature) and their responsibility to promote well-being. They are motivated by mutual benefit, environmental protection and well-being. Social capital is stronger and social solidarity underpins policy and actions. Linking social solidarity and circular actions has been successfully adopted in Paris. In contrast, societies with a low humane orientation are motivated by power and possessions. In these societies material wealth gives status and social capital is under-developed. The preference for consuming new (and unlimited) resources combines with low adaptive capacity (due to weak social capital) and a lack of concern for the environment. This cultural orientation would make it very difficult to implement circular development.

Culture also influences how society values circular development. Reused or recycled products are often undervalued, because they are perceived to be lower quality, with low performance, safety and health standards. For example, there is public opposition to the reuse of grey-water, because of the perceived risk to public health (Wilcox et al., 2016). Sometimes the distrust arises from past experience, which is then very difficult to shift. In Amsterdam, there is public distrust in grey-water systems because of a historical occurrence of cross-contamination, a valid concern arising from previous experience. This might only be overcome through successful trials and monitoring if it were reintroduced.

Refurbished buildings are also under-valued. Investors, owners and tenants are concerned about building performance (commercial and operational), maintenance costs, risk and uncertainty associated with older building stock (Bullen and Love, 2010). Building certification systems providing information about quality and performance of buildings, components and materials could help to generate demand for refurbished buildings. Labelling systems for second-hand and refurbished goods (as in Paris) and construction materials (material passports in Amsterdam) have also proved effective in generating demand. Thus, providing information may change values. In other cases, there are cultural taboos which would be difficult to overcome. For example, the recycling of sewage to produce protein for feedstock (as in Power to Protein, Amsterdam) maybe culturally unacceptable. However, no such taboos exist when converting sewage sludge into building materials (e.g. Thames Water, London) or biogas (e.g. ecocycles, Stockholm).

Green spaces and water are culturally valued, usually for their aesthetic appeal and recreational use. There is evidence to support that property prices increase with proximity to green spaces and water (De Groot et al., 2013; Okvat and Zautra, 2011; Roy et al., 2012). However, the majority of ecosystem services offered by the blue-green infrastructure appear to be culturally under-valued (in part because they are not understood). There is a public preference for well-managed and maintained blue-green infrastructure. The value falls if the green spaces and waterways are unkempt. Yet the more natural the habitat, the greater the biodiversity. The use of a green index in both Stockholm Royal Seaport and QEOP has helped professionals designing schemes understand the ecosystem service benefits of including green infrastructure into new developments. This has helped to inform the practices of those involved in implementing development.

A lack of connection between those living in cities and the natural environment (Trevors and Saier, 2010) reinforces this current values system. In addition, poor understanding of resource cycles reduces willingness to change (De Flander, 2015). Greater engagement of citizens in the stewardship of the urban ecosystem could help to increase a deeper cultural understanding of the benefits ecosystem services provide. For example, environmental stewardship and a cultural reconnection to local environments has been cultivated in Paris, Amsterdam and London through the urban agriculture movement.

Changing lifestyles and social practices

The local cultural values and systems of provision operating in a city interact, influencing the lifestyles and social practices of those inhabiting the city. Peoples' values and systems of provision influence their ability and willingness to engage in reuse and recycling practices. For example, community composting schemes and organic waste collection systems have been introduced in Paris to promote recycling behaviours. However, these systems of provision will not be effective if citizens place a low value on the importance of recycling food waste. Of course regulatory incentives can be imposed to address this.

Engagement in circular practices will also be affected by lifestyle and individual attributes. Engaging in circular practices requires time, money, energy, physical capacity, skills and expertise. For example, if waste separation or transferring compost to community facilities is too time-consuming (incompatible with lifestyle) or physically demanding (lack capability), recycling practices will not change. Generally, convenient and low cost, circular practices are preferred.

Circular practices which offer multiple benefits (i.e. are cost saving, provide an income, offer opportunities to socialise, provide health benefits) can also lower thresholds for engagement. For example, the community renewable energy projects in Brixton addressed fuel poverty (reduced costs) and provided an income for those involved. The skills and expertise needed to initiate and manage the scheme were provided by Repower London. Thus, motivations for engaging with the scheme were greater than the transactional costs, and so it has proved successful.

Many circular activities rely on volunteers (e.g. Brixton Pound Café). Yet engagement is time and energy consuming and premises are expensive. Thus, the transactional costs are high, making it hard to sustain circular activities. Subsidisation, particularly of those schemes with a solidarity focus (often the least well-funded), maybe needed in order for circular activities to sustain and ensure the knowledge generated and social practices developed aren't lost.

Passive systems which require minimal engagement from users reduce transactional costs. For example, the closed-loop system in Hammarby is a passive system. However, it requires users to operate the waste disposal facilities effectively. This has minimal transaction cost. Nevertheless, there is still misuse of the system provided (Williams, 2019b). Post-occupancy training in the effective use of systems appears to have had a limited effect (ibid). This suggests that a lack of awareness is not the problem. It is more likely connected to lifestyles and values.

Finally, cities bring together people from a variety of cultures. Cultural diversity will impact on the values, social practices and lifestyles adopted by those living in cities. It can also create variability in the success of adopting pro-environmental behaviours. Thus, a pluralistic approach will be needed to encourage the adoption of circular practices, services and products in cities.

Building social capital

Strong social capital[1] increases a community's capacity to adopt environmental praxis and engage in environmental stewardship (Williams, 2005). It increases the capacity of a community to react to events, collaborate, share resources (expertise, skills, financial, etc.) and learn from each other through strong social networks. It helps to build adaptive capacity in communities. Thus, it is fundamental to the successful implementation of a circular development. Engagement in circular activities can also help to build social capital. Thus, it creates a cycle of positive reinforcement. The challenge is how to build social capital in cultures based largely on individualistic values.

Social capital is greater nowadays amongst interest groups rather than within local communities. However, where local social capital is built in neighbourhoods, through communities of interest (e.g. Brixton Remade, Brixton Energy), collective actions are more easily enabled. This enables activities which promote ecological regeneration (e.g. urban farming). Self-identifying communities of interest sustain for longer (Williams, 2005). Common norms and values are shared within these communities. This enables trust to be built between individuals and encourages reciprocal relationships to develop, resulting in greater engagement in community activities (Metzler 2000; Williams, 2005). This creates the networks which are fundamental to resource looping (urban symbiosis).

It also enables learning within the communities and to an extent a degree of self-policing (ibid). Where social capital is strong, the burden of engagement (in terms of time, energy and finance) can be shared across a community. Thus, communities are able to adapt to contextual change. Building social capital within communities which value resources and ecosystem services will help to implement circular development.

Sharing and localisation have a fundamental role in the building of social capital. Local symbiotic capital (Chapter 2) can develop in communities (Curtis, 2003). For example, in Brixton strong social capital generated by the Transition Towns movement (and associated circular projects) was reinforced through local financial capital (local economic system and currency) producing natural capital (enhanced ecosystem services) and human capital (increase in diversity of local skills).

Equally sharing actions can help to build social capital in communities (e.g. sharing cars, lifts, goods, expertise, spaces) which enable more collective responses and environmental praxis to develop. Infrastructure provided in urban environments may help to build social capital (e.g. co-working and co-living spaces). Cohousing, which engages residents in financing, designing and managing new housing developments, has been extremely successful in building social capital and enabling environmental praxis (Williams, 2005).

New institutional arrangements, enabling community ownership and co-provision of assets help to build local social capital (e.g. social enterprises such as Transition Town Brixton). It is also important to prevent the loss of local social capital and circular expertise within communities, possibly through subsidies (e.g. payment of unpaid volunteers or provision of no/low cost premises) to ensure the transformation is sustained. Linked to this is the importance of keeping communities together. This can be threatened by the financialisation of housing and land speculation.

Cities have transient populations (e.g. students, economic migrants, digital nomads). Their mobility (and potentially limited community engagement) can weaken local symbiotic capital and threaten circular praxis. This is particularly an issue in global cities. Of course, transience does not negate environmental praxis or engage with local communities. However, the limited connection with place can make this shift more challenging. It is important to encourage these groups to engage with the local community and circular activities. This may be partially enabled through co-living and co-working spaces, which have helped to build social capital within some of these more transient communities.

It is also more difficult to encourage circular practices in new communities (e.g. Hammarby). Of the new developments presented in the book, De Ceuvel provides the best example. There were various success factors. A small number of actors were involved in the project. The group had common objectives (rules and norms) which aligned with delivering circular development (reinforced by tenders). The group was self-selecting. All actors were engaged in the design, planning and financing of the project from an early stage. Thus, they had a vested interest in its success. All actors wished to learn from the project, and have created networks through which their knowledge can be disseminated. For some this knowledge had commercial value too. Such an approach is possible for a relatively small-scale, non-residential project like De Ceuvel. However, it is more difficult to replicate for large-scale speculative developments (e.g. Hammarby, SRSP, QEOP and Clichy Batignolles).

Knowledge creation and smart data

Information is critical for the transformation of values, practices and institutions needed to support circular activities. A variety of capacity building programmes have been developed for urban actors engaged in circular development. Amsterdam, London and Paris use urban experiments to test new business models and development trajectories. Discussion forums have been set up in London (Circular Economy Club), Paris (*Paris Esprit d'Entreprise*) and Amsterdam to learn from participant experience and share knowledge. Skills workshops for infrastructure providers and businesses have been established in Amsterdam, Paris and London to encourage circular development. Engagement in international networks (*CE100, Eurocities, C40 cities*) ensures that knowledge created is spread more widely. However, more data is needed to identify the benefits of adopting circular development.

In the age of smart cities and big data, the amount of information collected is increasing. Data can be used to raise awareness of resource flows in cities.

Information helps to tackle limited public awareness and understanding of resource cycles (water, nutrients and materials), ecosystem services, the future impacts of major events and climate change (Darby, 2006; Fischer, 2008; Ueno et al., 2006). For example, the online data platform *REFLOW* in Stockholm visualises hidden resource flows in the city and is used as an educational tool to encourage looping practices.

Data also assists urban actors in planning for future catastrophic events. For example, apps draw on a range of data (satellite data, social media, crowd-sourced data, historical data, meteorological data) to predict disastrous events and allow urban populations to prepare for them (e.g. *I-React*). Other apps allow professionals to identify the best adaptation measures for the local context they are working in (e.g. *Climate Adaptation App*). Thus, smart data can help build adaptive capacity in cities.

Collecting comprehensive, consistent, useful data in cities is a major challenge. Civil society may provide one solution. Increasing engagement of civil society in the generation of data (crowd-sourced data) can produce smarter communities, who actively participate in the circular transformation process. For example, in Amsterdam, Paris and London, data platforms (populated by user groups) enable the exchange of soil (e.g. *Sol-Dating*) and concrete (e.g. *Sustainability Cloud*). They also identify unoccupied sites and buildings for reuse (e.g. *L'Association Plateau Urban*) and urban mining opportunities (*PUMA*). Currently, the engagement of civil society in generating ecosystem service data for cities is limited. More information on urban biodiversity is needed to understand local, city and regional scale relationships between biodiversity and the generation of urban ecosystem services (Anderson et al., 2017; McPhearson et al., 2016). Thus, the role of smart data in the protection of ecosystem services in cities remains under-developed.

Data platforms can also enable engagement in circular activities. For example, in Paris, there are online platforms for locating repair services (e.g. *oureparer.com*) and enabling waste-free shopping (e.g. *LOOP*). In Amsterdam there are apps for identifying left-over food waste (e.g. *"Too good to go"*). Product labelling systems operating in Paris (for electronic goods, furniture and textiles) and Amsterdam (for construction waste) enable the reuse of a range of materials and goods. The online lifecycle analysis tool used in Stockholm supports contractors in reducing construction waste generated on building sites. However, there is a challenge to limit the number and diversity of platforms which are available, in order to enable a degree of coordination.

Data provides urban politicians and managers with the technical evidence that transformation is needed and that regulation is effective (Anderson et al., 2017; Bullen and Love, 2010; Lacovidou and Purnell, 2016). However, data monitoring urban metabolism has been collected for only a few cities worldwide. Examples include the *Urban Metabolism Platform* in Paris and *Circle Scan* in Amsterdam. Interpretation issues exist due to a lack of common conventions (Browne et al., 2009; Kennedy et al., 2007; Zhang, 2013). Most urban metabolism studies use highly aggregated data – often at the city or regional level – that provides a snapshot of resource or energy use, but no correlation to locations, activities or people (Pincetl et al., 2012). There is a high data requirement for monitoring resource flows, a lack of follow-up and evaluation of the evolution of a city's urban metabolism and difficulties in identifying cause-and-effect relationships of the metabolic flows (Shahrokni et al., 2014). Data for vacant land and buildings is generally collected separately. Similarly, it is highly aggregated and often incomplete, as exemplified by the London *Brownfield Sites Review* (ARUP, 2018).

Issues around data ownership, privacy and commercial competitiveness restrict access to urban data (Ehrenfeld and Chertow, 2002; Herold and Hertzog, 2015; Khan et al., 2014). The quality of the data produced is also a concern due to limited coverage, inconsistent monitoring and frameworks (Allwinkle and Cruickshank, 2011; Lacovidou and Purnell, 2016). This reduces trust in the information exchanged (Lenhart et al., 2015). The platforms (virtual or non-virtual) for communicating and sharing data can be useful, but are highly dependent on the quality of the data they provide. Certainly, a lack of both can create a real challenge for circular activities (Boons et al., 2011; Ehrenfeld and Chertow, 2002). Thus, monitoring and managing urban resource flows and ecosystems services is difficult.

Reviewing the regulation

The European policy framework is well developed and supportive of looping actions. Circular economy is the focus for the vision for a competitive Europe (EMF et al., 2015). This is further supported by the *Europe 2020 Strategy*, the *Roadmap for a Resource Efficient Europe initiative, the Zero Waste Programme for Europe and Closing the Loop: an EU Action Plan for Circular Economy.* These plans and strategies have led to the production of circular economy strategies for France and the Netherlands. For the UK, circular economy is embedded in the industrial strategy (and thus has a rather narrow focus). For Sweden, resource looping has been reinforced by its voluntary commitments to the *Natural Step* and *Aalborg Charter.*

The primary legislation influencing resources at an international level remains sector specific (e.g. the *Water Framework Directive, Urban Waste Water Directive, Renewable Energy Directive, Energy Efficiency in Buildings Directive, Waste Framework Directive*) rather than integrative. This is often reflected in national legislation. At a local level this regulatory framework tends to reinforce siloed-thinking and sector-specific strategies for managing resources in cities. This creates a barrier to cross-sectoral looping actions and nexus solutions. A key challenge is to create joined-up, cross-sector regulation which is supportive of looping actions.

Regulatory standards can be a useful tool for ensuring quality both in the production and performance of looped resources. This provides certainty for regulators, investors and consumers. For example, the adoption of a publically visible standard with proven credentials has supported improvement in the public perception of grey-water reuse and helped systems scale-up in cities (Wilcox et al., 2016). Equally standards set for urban mining have helped enable repurposing, recycling and reuse of materials and infrastructure in cities (Bastein et al., 2017; Lacovidou and Purnell, 2016; Ortner et al., 2014). However, standards can create a barrier to looping actions. For example, building regulations and conservation standards create regulatory barriers to adaptive reuse of infrastructure (Bullen and Love, 2010). So the challenge is to create a set of standards which indicate the quality of looped resources. This will also help to establish greater economic and cultural value for these resources.

The regulatory framework relating to ecological regeneration in member states is encompassed by the *EU 2020 Biodiversity Strategy*. However, the strategy does not focus on biodiversity in cities. The Strategy requires that "by 2020, ecosystems and their services are maintained and enhanced by establishing green infrastructure and restoring at least 15% of degraded ecosystems". It recognises the synergies between green infrastructure and environmental policies such as land use, air quality and water. Several other EU policies and the *Green Infrastructure Strategy* are working to integrate green infrastructure and harness its potential for agriculture and forestry, climate change mitigation and adaptation and disaster prevention. However, there is no urban focus which encourages green infrastructure provision as a mechanism for improving ecosystem services in cities.

There is a clear regulatory framework for climate mitigation driven by the legally binding Paris targets, agreed through the *UN Convention on Climate Change.* These cover all sectors of the economy, aiming to achieve at least 40% greenhouse gas emissions reductions by 2030. This legislative framework provides investors with a clear and predictable vision of the way forward. It addresses both structural elements to underpin climate and energy action, and specific provisions for sectoral action where it was needed. The EU is the first major economy to fully implement its Paris Agreement commitments in legally binding domestic laws. The UK was the first member state to adopt legally binding targets for reductions and carbon budgeting through the *Climate Change Act in 2008.* Links are beginning to be made between circular economy and greenhouse gas emissions (EMF, 2019), which may increase regulatory support for looping actions (via mandatory mitigation targets).

However, the regulatory support for adaptation is less. In 2013 the European Union adopted a *Strategy for Adaptation to Climate Change.* It encouraged member states to produce climate adaptation plans and many have. It also supports climate adaptation (and mitigation) in cities through the *Covenant of Mayors for Climate and Energy Initiative.* However, this is a voluntary initiative. Thus, there is no mandatory regulatory framework underpinning measures for climate adaptation in cities. However, there are relevant directives covering flooding, air pollution and water. Thus, the challenge is to create stronger regulatory framework (reinforced by directives), supporting all three circular actions and recognising the synergies between them, in cities.

Many layers of regulation – primary legislation (protocols and treaties; directives and acts) and statutory instruments (building codes, ordinances, contractual agreements and plans) affecting resources, ecosystem services and resilience – coalesce in cities. A range of non-statutory instruments – targets, strategies, policy frameworks – also guide development in cities. The Pact of Amsterdam adopted in 2016 sought to strengthen the urban dimension of EU policies to coordinate EU policies impacting cities and involve cities in EU policy developments. It resulted in the creation of the *Urban Agenda* which aimed to streamline international, national and local regulation in cities to reduce regulatory conflicts and create greater consistency. It focusses in various priority areas including circular economy (looping), sustainable land use (ecological regeneration) and climate adaptation. Streamlining these regulatory frameworks remains a significant challenge.

Cities have various statutory and non-statutory instruments, which they can use for delivering circular actions. Cities use targets to signal their future vision and outline pathways to achieving those targets. The most common targets, at least across the cities we have considered, relate to waste (e.g. zero waste, Paris and London) and greenhouse gas emissions (e.g. zero carbon, London). One challenge would be to produce a set of targets which helped to encourage all three circular actions, without creating conflicts. Cities also produce strategies for achieving circular economy (Paris and London) or incorporate circular economy into their other strategic plans (Amsterdam and Stockholm). These outline policy goals, strategies and instruments for getting there. However, only in the Paris *Circular Economy Plan* are clear linkages with adaptation and ecological regeneration made.

The spatial development plan – a statutory instrument – can bring together all three circular actions in a city through the development process. In Amsterdam, the spatial development plan has policies to encourage high-value reuse of buildings, components and materials; the construction of "smart" adaptive buildings and blue-green infrastructure. However, in some plans (e.g. the London Plan) these activities remain separate; thus, the synergistic benefits of adopting looping, adaptive and regenerative strategies together are overlooked. The spatial plan can also be used to ensure urban form and infrastructure support circular activities. For example, in Stockholm, the spatial plan has supported *ecocycles* by encouraging high density, mixed-use development, alongside the expansion of the district heating and public transport systems (Williams, 2013).

Circular development can also be encouraged through a process of land issue and tendering; the use of environmental performance programmes or application of planning conditions to new development. In Amsterdam a circular land issue and tendering encouraged all three circular actions. However, implementing circular tendering proved challenging. Developers, construction, demolition, disposal companies and planning authorities lacked the expertise to deliver or oversee circular projects. Capacity needs to be built amongst key actors to enable the process.

Clear guidelines for what constitutes circular development; procedures for design, procurement, construction and disposal; and designation of responsibilities for implementation are needed. Tools such as the *green space index* and *Lifecycle*

Assessment Analysis used in London and Stockholm could prove useful in creating circular solutions. However, this will also lengthen development timelines and require additional expertise. Lessons from Hammarby suggest that post-occupancy monitoring and enforcement will be required to ensure the goals set by the circular tendering process have been achieved in practice.

Another challenge is to streamline the regulatory framework to enable circular development. De Ceuvel highlighted the need to review planning controls to enable circular development. Similarly, restrictive conditions are placed on the reuse of buildings and grey-water recycling; neighbouring land-use or buildings for multiple uses (due to incompatibility between uses); green and blue infrastructure (perhaps because of the risk they might pose to person or property) often due to health and safety concerns. The problem lies in the prescriptive nature of the guidelines for addressing health and safety considerations. Moving to a performance-based approach, which requires all circular development to be safe, but allows some flexibility in the way this is achieved, is likely to be more successful. In Paris, urban experiments are being used to explore legal barriers to circular development, to enable this streamlining process to begin.

Political challenges

A global shift towards neoliberalism has significantly affected the political framework in which European cities operate. Neoliberalism is associated with laissez-faire economic liberalism and free-market capitalism. Neoliberal policies include privatisation, deregulation, globalisation and reduction in public spending. These policies aim to increase the role of the private sector in the economy and society and reduce the role of the state. Neoliberalism is based on neoclassical economic understandings, and thus does not adequately value or protect resources or ecosystem services. Consequently, neoliberal politics is likely to be in conflict with implementing the circular development, which is an enormous challenge to overcome.

Neoliberalism has influenced policies, instruments and funding decisions in European cities. It has changed the number and diversity of actors involved in service and infrastructure provision; altered power relations between key actors; and shifted the municipalities towards a more facilitative role in urban governance (Williams, 2016). It has also resulted in a reduction in public funding for new development (infrastructural projects) and the privatisation of services (waste, water, energy, transport, etc.). Thus, power has shifted away from local government towards the private sector. Private sector actors tend to prioritise economic goals. They also tend to be more risk averse and less likely to invest in innovation than their public sector counter-parts (Mazzucato, 2011). Thus, circular transformation is extremely challenging in this context.

Globalisation has also increased the percentage of international actors rather than local actors involved in systems of provision in cities. This further reduces the control local government has over the delivery of infrastructure and services.

International actors are likely to be guided by international regulation and economic interests, rather than local, social and environmental interests. Without a strong, supportive regulatory framework, or economic incentives, there is limited market incentive for private actors to protect resources and ecosystem services.

In a neoliberal context "state interference" (regulation and economic incentives) is minimised, and thus there are limited levers for transformation. Neoliberal systems use network governance rather than command and control government to align goals and deliver development. Network governance relies on capacity building to create networks and reciprocity between actors. It is local government's responsibility to build both. Thus, in the neoliberal context, the role of local government has shifted from one focussed on provision and regulation towards enabling and self-governance (Bulkeley and Kern, 2006). Cities implementing circular development largely rely on enabling tools (e.g. capacity building workshops, skills training, network creation, online skills and networking platforms) and public procurement to deliver the transformation (e.g. exemplified by the London's approach to circular economy). This approach may produce pockets of innovation, but rarely systemic transformation.

To implement a circular development pathway will require radical economic, cultural, institutional and technological restructuring. Such a transformation will require long-term political support and leadership. This is not encouraged by the current political culture of short-term, market-driven, reactive decision-making. Existing cultural values and short-term political cycles underpin this present-orientated view. Thus, it will be hard to shift. However, in order to create policies that support futurity and inter-generational equity, resource protection, ecological regeneration and community adaptiveness, the underlying political culture and systems of operation will need to be addressed. Stockholm provides an example where long-term political support (and public funding) for closed-loop systems and ecological regeneration have produced such a transformation.

Political priorities may sometimes conflict with circular actions and vary between national and local levels of government. For example, in the UK national political support for foreign and corporate investment in property and land markets has prevented the reuse of vacant property and use of land for industrial activities in London. The national economy is very dependent on this investment and thus it is a national political priority to support it. However, at a local level addressing the lack of affordable accommodation, the creation of local jobs and diversification of the economic base in London are higher priority. Of course, this creates conflicting priorities. It also prevents the reuse of infrastructure and allocation of space in the city for low value circular activities. The challenge here is to create political motivation for prioritising circular development goals (Figure 8.1).

In addition, the lack of an integrated approach to policy-making at a local level may also create barriers to the delivery of circular development. Linking circular economy with ecological regeneration and adaptation, as is the case in Paris, helps to maximise the benefits which come from the synergies between these actions. It also expands the potential benefits which accrue from adopting the policies (the win-wins). For example, the policy to reuse food waste in Paris reduces landfill and associated greenhouse gas emissions, whilst redistributing food to those who need it. It has also spawned new businesses (food reuse eateries). In contrast, policies which present the circular actions separately do not benefit from acknowledging the synergies between them (as is the case in London). The challenge is to encourage policy-makers to take a more systemic approach in their policy-making, recognising the synergies particularly between circular actions.

Institutional reform

The current cultural values, economic, political systems and legislative framework greatly influence institutional culture, structures and practices in our cities. Thus, neoliberal politics reduces the role of local state institutions in the circular transformation process, placing the responsibility with private (often global) institutions. Many of these institutions are risk averse and profit-orientated, which creates path dependency and technological lock-in.

Private institutions are poor at engaging non-state actors in projects and less likely to deliver public benefits than public bodies, unless under a service-based contract (Da Cruz and Marques, 2012; Furlong and Bakker, 2010). In some instances, the global nature of private institutions means they are culturally different, have different priorities and are governed by legislative frameworks operating in other spaces. These institutions rely on the privatisation of assets for revenue, which results in difficulties sharing data and resources.

Many institutions are embedded in neoclassical economic thinking. Thus, they undervalue natural resources and ecosystem services. Economic goals are prioritised. Culturally, there is a bias towards short-termism, individualism and materialistic goals, all of which cut against the circular development pathway. Sectoral and professional segregation reinforced by the sector-specific legislative frameworks prevents the integrated, systemic approach needed to encourage circular systems (Roelich et al., 2015; Smith, 2007). In combination, these create major challenges to the emergence of institutions capable of supporting circular development.

Institutional reformation will be required to deliver circular development, although some institutions are already well suited to this paradigm. However, this will also require shifts in economic models, political and cultural values, for example, a shift away from neoclassical economic models to ecological ones. Institutions will need to take a long-term view and invest for the long-term. Investment decisions should take account the societal value of resources and ecosystem services. Institutions will need to take risks, innovate and promote internal cultures enabling this. There will need to be a shift from resource-based to service-based, linear to circular business models. Institutions should empower communities, encourage co-provision and community ownership and create prosumers and environmental stewards. Data should be created and owned by citizens. Instead of actively encouraging individualism as a means of increasing consumption, institutions should support a more collective response to societal problems.

There should also be a shift towards institutions which have greater local responsibility, to whom the local community and environment matter. This could just be a matter of increasing local government intervention, enabling local government to use all of the competencies and levers at their disposal to enable transformation. However, it will also require local government works with and empowers local communities. This will enable communities to take responsibility for the changes taking place. It will mean an expansion in institutions such as cooperatives (e.g. Brixton Transition Town Food Coop), community companies (e.g. Brixton Energy) and social enterprises (e.g. Brixton Transition Town; Our Park Life, Queen Elizabeth Olympic Park). These institutions will also need to become more agile, resilient and able to change with the shifting context. It will involve the growth of more comprehensive interconnectedness, social networks and social learning in cities.

Institutional capacity will need to be built to support circular actions in cities. There will need to be new institutions to produce and enforce standards for recycled and reused resources and adaptable infrastructure; to support new ownership models which allow the reuse of goods and infrastructure (Bastein et al., 2013); to regenerate ecosystem services in cities and monitor the impact and to collect, share, monitor and regulate the use of data needed to encourage recycling of resources (Allwinkle and Cruickshank, 2011; Townsend, 2013). New institutions will be required to support learning within industry, commerce and the community in order to change systems of provision, social practices and lifestyles which undermine circular actions (Barragán-Escandón et al., 2014; Bullen and Love, 2010). Of course there will be institutional (cultural and structural) inertia to change because of vested interests in preserving current practices and minimising risk. These will need to be overcome to facilitate circular actions.

Ecological challenges

The local environment may challenge the implementation of the circular development. The extent to which the urban ecosystem is degraded can affect its capacity to ecologically regenerate and for resources to be looped. For example, land contamination reduces the potential for grey-water reuse and soil recycling (Bullen and Love, 2010; Wilcox et al., 2016). It also inhibits ecosystem services including provisioning services (e.g. urban agriculture and forestry) and regulating services (water storage). Lack of vegetation will influence the ecosystems' capacity to absorb carbon-dioxide, reduce air pollutants, regulate temperature and store water. These problems will be exacerbated by local climate, relief, topography and hydrology. Thus, circular development pathways will need to appropriate for the local environment in which they are implemented. This is well illustrated by the De Ceuvel case. Water, energy and building systems were designed to work with high levels of soil contamination, whilst using phytoremediation to decontaminate the soil.

De Ceuvel also highlights a second challenge. Ecological regeneration using biological techniques takes longer than more traditional, non-biological approaches. This can be problematic to manage in short political cycles (Anderson and Minor, 2017). Longer-term political support for this approach would be needed to ensure continuity and to secure the funds for the ongoing management of the biological processes. Taking a biological approach may also slow the development process. This may be problematic in land scarce environments, particularly where there is pressure for development. However, temporary permissions for activities which promote ecological regeneration may be popular amongst land owners, as long as the activity provides revenue in the short-term. In the long-term the value of the ecologically regenerated site increases.

A further challenge is presented by the ongoing management and maintenance of blue-green infrastructure. This has usually been the responsibility of the local public authority. However, increasingly this work has been contracted out to private actors. Yet due to the under-valuing of ecosystem services, the financial reward is often insufficient. In some new developments there has been a move towards encouraging residents to take responsibility for blue-green infrastructure within their neighbourhoods (e.g. in Stockholm Royal Seaport and QEOP). This approach is in part to reduce the costs to the state, which is particularly popular during a period of public funding austerity (and fits with the neoliberal model). However, public engagement in the stewardship of the local environment has other benefits too. It helps to address the lack of public connectedness with the local environment. This raises public awareness of environmental problems and solutions, whilst building the expertise and skills needed for environmental stewardship.

The public could be engaged in a range of activities (e.g. community renewable energy generation; conservation of biodiversity, greenways, water ways; gardening, urban agriculture or forestry). Their engagement in these activities will help to regenerate the urban ecosystem services and build social capital which enables collective action. In combination these actions will also increase the adaptive capacity of communities. The major challenge is how to get people to engage in environmental stewardship in their city. There are many transaction costs which may prevent engagement. Thus, financial resources to enable environmental stewardship programmes are needed. Workshops to increase citizens' expertise or engage experts to work with communities to help them to manage and maintain urban ecosystems will be essential.

Allocating time during the week for these activities is also important. Perhaps, this could be achieved by integrating them into the school curriculum or working week. Alternatively shortening the working/school week could help to enable engagement in circular activities. Motivation to be engaged in schemes might come from positive outcomes, for example, the financial rewards or opportunities to socialise. It equally could be driven by the desire to react to local (e.g. flooding, drought, air pollution) and global environmental problems (e.g. climate change). Reward systems, for example, personal carbon allowances, could also motivate engagement. Of course monitoring the impact of the stewardship programmes could provide motivation, as well as encourage further public investment.

Technical and design challenges

Circular design and thinking has not been incorporated into urban systems. Systems are linear, segregated and use grey infrastructure (Unruh, 2000; Williams, 2016). These systems are locked-in by the vested interests and sunk cost of those providing them. Citizens also develop lifestyles and social practices which fit with grey infrastructure and linear systems. This creates a socio-technical lock-in, which reinforces linear and separated systems of provision, which impedes the implementation of circular solutions.

Even if there is willingness amongst providers to adopt circular systems of provision, it is practically difficult to alter infrastructural systems due to the capital cost and disruption generated by such a radical transformation. New infrastructural systems are often integrated into new developments, when the costs can be borne by the developer or consumer. However, there is limited opportunity for the renewal of infrastructural systems in most cities (certainly in Europe), as development rates are low.

Financing new technical systems can be a major challenge. Utilities prefer to off-load the cost of infrastructure provision onto developers (where possible) creating off-grid alternatives. This was demonstrated in QEOP by the grey-water recycling system. This approach does not support wider community transformations, but it means the existing system is not overloaded by the additional development and the cost is eventually borne by the homeowner.

The replacement of grey infrastructure with blue-green infrastructure may also create a financial challenge. Although it may be cheaper initially to implement in terms of capital cost, the overall operational costs may be higher. The maintenance of these systems requires new skills, which has an associated cost. Green infrastructure also takes longer than grey infrastructure to become established and to operate effectively. Finally designing redundancies in systems, to ensure they are adaptable, also has cost implications. Encouraging investors to consider the future value of these design decisions and technical systems is critical for successful implementation.

There are some separate design tools which enable the whole lifecycle of buildings to be compared; the impact of green infrastructural solutions to be calculated and adaptive design solutions to be tested in different climate scenarios. However, design tools which enable the integrated impact of looping, adaptive and regenerative infrastructural solutions to be compared, are needed. Ideally, this would also be linked to tools which could estimate the future value of the solutions generated, in order to build a case for specific design decisions.

Summary

Circular experiments are emerging at a variety of scales, across cities in Europe. However, there is little evidence as yet, that these experiments have transformed the development regime in which they are embedded. There are numerous challenges to transformation (Figure 9.1). The greatest challenges are political, economic and cultural. All three are intrinsically linked and in conflict with circular development.

The neoliberal political system prioritises the economy and relies on the market to deliver the circular transformation, which will benefit society. However, the economic system does not reward these benefits. Thus, there is no incentive for the market to do so. Existing cultural values – individualism, short-termism and materialism – prevalent in a neoliberal society also conflict with circular development. Development regimes which exist in future-oriented, collectivist societies, where socio-ecological benefits are valued, are more likely to transform.

This combination of challenges experienced in the neoliberal context makes it difficult for circular actions to be applied to urban systems, unless they are commercially viable (e.g. generating energy from waste). This is demonstrated in the most neoliberal case study – London. Here circular solidarity experiments (e.g. the Brixton Pound Café) and commercially non-viable schemes (e.g. the black-to-grey water recycling scheme in QEOP), which benefit society, cannot be sustained. Thus, the capital generated by these experiments is lost and the development regime remains unchanged.

Amsterdam provides a good example of where commercially viable circular systems have begun to emerge, based on biomass and construction waste. These systems are likely to alter the development regime in the future. However, they have benefited from significant political support (at a national and local level), public investment and regulatory support, which has enabled their development. Less commercially viable projects (e.g. De Ceuvel) have received limited supported on a temporary basis. It remains to be seen whether these experiments will persist and transform the development regime.

Stockholm illustrates how a circular transformation can occur in a more amenable political and cultural context, but can collapse as neoliberal values and policies take-hold. Culturally Swedes value their environment and have adopted a more collectivist approach to society. Ecocycles emerged during a period when a red-green coalition, with strong socio-ecological values, was in power in Stockholm. During this period circular principles were integrated into the thinking of development decision-makers and tested in practice. The more recent shift towards neoliberal politics has resulted in the privatisation of services and replaced local public actors with private (often international) actors. This shift has created difficulties in delivering circular development and ecocycles in Stockholm.

Paris also provides an interesting case. Although operating within a global, neoliberal context, the government still adopts an interventionist approach (using regulation and funding) to development. It also places value on socio-ecological

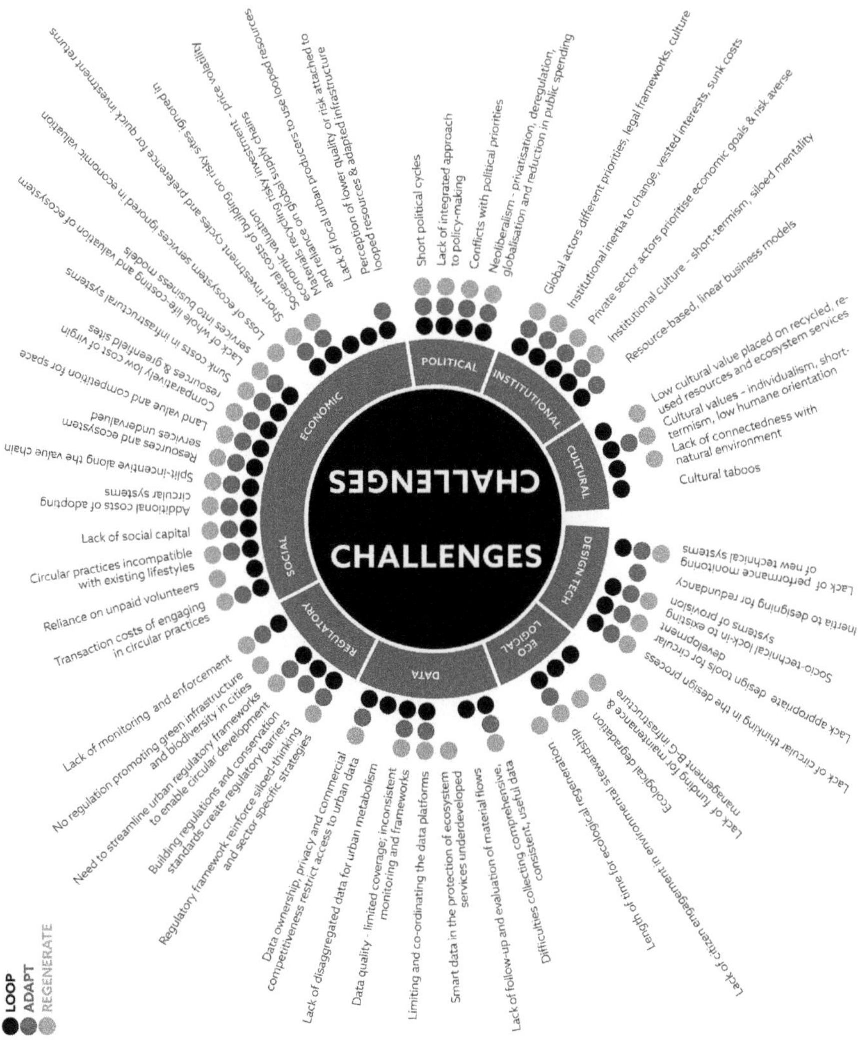

FIGURE 9.1 Challenges to circular development.
Source: Authors own produced by Draught Vision Ltd.

goals. Thus, it is promoting circular development with a socio-ecological focus (e.g. urban agriculture, food-reuse, buildings reuse, decarbonisation of the energy supply) through designation of sites, funding, procurement and the use of local provisioning powers. This has also been supported by regulation at a national level (e.g. supermarket food reuse legislation). Thus, in Paris there seems to be potential for these solidarity, circular experiments to modify the development regime. These findings suggest that even within a wider neoliberal context a circular transformation may be possible within a city, if there is the political will (Figure 9.1).

Note

1 Social capital is the effective functioning of social groups through interpersonal relationships, a shared sense of identity, a shared understanding, shared norms, shared values, trust, cooperation and reciprocity.

10

CIRCULAR CITIES

Conclusions and future research

What is a circular city and circular development?

To date, the focus in the academic literature has been on circular economy and encouraging the emergence of circular businesses in cities. In this book we have moved beyond this, towards a socio-ecological conceptualisation of circular cities and circular development (Chapter 2). Circular cities are urban systems in which resources are looped, the ecosystem is regenerated and the socio-technical systems (infrastructure and communities) evolve with changing contexts. Thus, circular cities are resource efficient, resilient and operate within the global carrying capacity. Circular development implements circular systems, activities and infrastructure in cities, largely through the spatial planning and economic development processes. It has the potential to address several sustainable development goals often overlooked: inter-generational equity, futurity and environmental protection. If adopted by the development regime it may truly offer a revolution in urban sustainability.

The European case studies illustrate how circular cities and circular development might manifest in practice, across a range of contexts (Chapters 3–6). There is great variation even amongst these few case studies. However, there are also commonalities between cities. There is a common focus on four resource flows: organic waste, construction waste, grey- and waste-water. These require common infrastructural solutions: waste-to-energy systems; grey-water recycling systems; green infrastructure, adaptable infrastructure. The cities adopt similar circular activities (e.g. food reuse, composting, urban farming) and levers for transformation (regulation, capacity building, procurement and provisioning). These findings begin to demarcate common features of circular cities. A series of development pathways have begun to emerge from the case studies. These form the basis for a typology of circular development pathways (Chapter 7). So far three types have been identified. It is likely that other pathways will emerge, as we explore more cities across Europe. Thus the typology will develop.

Dynamics between circular actions and other urban strategies

The case studies demonstrate the synergies between circular actions (Chapter 7). They underline the importance of implementing circular actions in combination, thus reinforcing the conceptualisation of circular development, presented in Chapter 2. The case studies also demonstrate the synergistic relationships and conflicts between circular actions and other urban strategies (localisation, substitution and optimisation). These dynamic relationships require further investigation to determine with certainty which strategies are likely to operate together successfully in circular cities.

Reasons for adopting a circular development pathway

The case studies confirm there are benefits accrued from adopting circular development in cities (Chapter 8). In addition to the ecological benefits, it creates healthier, more adaptive living environments, a more diverse economic base with a range of jobs, greater community engagement and social solidarity. However, these benefits are not equally shared across the population, thus producing social inequalities. The potential social disadvantages (especially to the urban poor) require further research. However, it is argued that these problems could be addressed, given the political will to do so (Chapter 8).

The benefits and disbenefits of circular development, both within a city and to the wider community, need to be better understood and quantified. Given the global imperative to tackle climate change, calculating the impact of adopting a variety of circular development pathways on greenhouse gas emissions should be prioritised. The potential benefits of adopting the approach in new contexts (e.g. cities in developing nations, those suffering from shrinkage or resource insecurities) should also be explored.

Challenges to circular urban transformation

There are major challenges to the circular transformation of the existing development regime, which in many instances requires a shift in political philosophy, cultural values and assumptions underpinning the economic system (Chapter 9). However, our case studies demonstrate that where there is the political will and the resources, circular transformation could be possible. However, circular experiments need long-term, political and financial assistance and a supportive regulatory framework to ensure they scale-up (Williams, 2016).

Through the process of cosmopolitanisation the ideas mobilised in European cities could translate into new contexts (Williams, 2017). For example, in countries where political, economic systems and cultural values are more aligned with circular development, circular cities may flourish. Equally in urban systems which might benefit from adopting this approach, for example, shrinking cities and resource insecure cities, the development regime may transform. This will very much depend on the benefits of adopting circular development being made clear to decision-makers. More research is needed to determine the challenges to circular urban transformations in non-European contexts.

BIBLIOGRAPHY

AEB Amsterdam, 2015. *For a Clean Society*. AEB: Amsterdam, The Netherlands.

Agudelo-Vera, C.M., Mels, A.R., Keesman, K.J. and Rijnaarts, H.H., 2011. Resource management as a key factor for sustainable urban planning. *Journal of Environmental Management*, 92(10), pp. 2295–2303.

Allwinkle, S., Cruickshank, P., 2011. Creating smarter cities: An overview. *Journal of Urban Technology*, 18, pp. 1–16. doi:10.1080/10630732.2011.601103.

Amsterdam Economic Board, 2018. Organic waste: The start of something beautiful, https://amsterdameconomicboard.com/en/nieuws/groente-fruit-en-tuinafval-het-begin-van-iets-moois (accessed 24-04-20).

Amsterdam Metropolitan Region, 2016. Agenda for spatial planning and the economy in the Amsterdam metropolitan region, Report published by the AMR.

Anderson, E.C. and Minor, E.S., 2017. Vacant lots: An underexplored resource for ecological and social benefits in cities. *Urban Forestry & Urban Greening*, 21, pp. 146–152. doi:10.1016/j.ufug.2016.11.015.

ARUP, 2018. The UK Brownfield sites review. https://www.arup.com/projects/brownfield-sites-review (accessed 10-08-18).

Atkins Global, 2012. Enabling the Olympic Park, Atkins Website. https://www.atkinsglobal.com/en-gb/media-centre/features/enabling-olympic-park (accessed 24-04-20).

Balseviciene, B., Sinkariova, L., Grazuleviciene, R., Andrusaityte, S., Uzdanaviciute, I., Dedele, A. and Nieuwenhuijsen, M.J., 2014. Impact of residential greenness on preschool children's emotional and behavioral problems. *International Journal of Environmental Research and Public Health*, 11, pp. 6757–6770.

Barragán-Escandón, A., Terrados-Cepeda, J. and Zalamea-León, E., 2017. The role of renewable energy in the promotion of circular urban metabolism. *Sustainability*, 9, p. 2341. doi:10.3390/su9122341.

Bastein, A.G.T.M., Roelofs, E., Rietveld, E. and Hoogendoorn, A., 2013. *Opportunities for a circular economy in the Netherlands* (pp. 1–13). Delft: TNO.

Bastein, A.G.T.M., Verstraeten-Jochemsen, J.N., Rietveld, E., Hauck, M., Frijters, E., Klijn, O. and Driessen, B., 2016. *Circular Amsterdam. A vision and action agenda for the city and metropolitan area.* TNO.

Bilitewski, B., 2007. Circular economy in Germany. In: Eleventh International Waste Management and Landfill Symposium, Cagliari, Italy, 1–5 October 2007.

Bishop, P. and Williams, L., 2012. *The temporary city*. London: Routledge.

Blok, K., Hoogzaad, J., Ramkumar, S., Ridley, A., Srivastav, P., Tan, I., Terlouw, W. and de Wit, M., 2017. Implementing circular economy globally makes Paris targets achievable, Published by *CIRCLE ECONOMY and ECOFYS*.

Bolund, P. and Hunhammar, S., 1999. Ecosystem services in urban areas. *Ecological Economics*, 29(2), pp. 293–301.

Boons, F., Spekkink, W. and Mouzakitis, Y., 2011. The dynamics of industrial symbiosis: A proposal for a conceptual framework based upon a comprehensive literature review. *Journal of Cleaner Production*, 19, pp. 905–911. doi:10.1016/j.jclepro.2011.01.003.

Bosman, F., 2011. Buiksloterham 2.0. Flexibele speelregels als randvoorwaarde voor success. *Plan Amsterdam*, 2, pp. 16–23.

Boulding, K. E. (1966). The economics of coming spaceship earth. In H. Jarret (Ed.), *Environmental quality in a growing economy*. Baltimore, MD: John Hopkins University Press.

Bowler, D., Buyung-Ali, L., Knight, T. and Pullin, A., 2010. Urban greening to cool towns and cities: A systematic review of empirical evidence. *Landscape and Urban Planning*, 97, pp. 147–155.

Brick, K., 2008. Follow up of environmental impact in Hammarby Sjöstad. Stockholm, Sweden: Grontmij AB.

Brinkley, C., Birch, E. and Keating, A., 2013. Feeding cities: Charting a research and practice agenda toward food security. *Journal of Agriculture, Food Systems, and Community Development*, 3(4), pp. 81–87.

Brixton Pound. http://brixtonpound.org/blog/2017/05/23/were-hiring-come-and-work-in-our-cafe/ (accessed 24-04-20).

Browne, D., O'Regan, B. and Moles, R., 2009. Assessment of total urban metabolism and metabolic inefficiency in an Irish city-region. *Waste Management*, 29, pp. 2765–2771.

Bulkeley, H. and Kern, K., 2006. Local government and the governing of climate change in Germany and the UK. *Urban Studies*, 43(12), pp. 2237–2259.

Bullen, P. and Love, P., 2010. The rhetoric of adaptive reuse or reality of demolition: Views from the field. *Cities*, 27, pp. 215–224. doi:10.1016/j.cities.2009.12.005.

Byström, J., 2018. The fifteen circular steps for cities. European Investment Bank Report.

Camaren, P. and Swilling, M., 2012. *Sustainable resource efficient cities: Making it happen*. Nairobi, Kenya: UNEP.

Carlisle City Council, 2011. *Carlisle—The Big Green City*. Penrith: Rebanks Consulting.

Cashmore, C., 2015. *Speculative vacancies 8: The empty properties ignored by statistics*. Australia: Prosper.

Chaparro, L. and Terradas, J., 2009. *Ecological services of an urban forest in Barcelona*. Barcelona: Centre de recerca ecoligica i aplicacions forestals, Universtitat untonma de Barcelona Bellaterra.

Chawla, L., Keena, K., Pevec, I. and Stanely, E., 2014. Green schoolyards as havens from stress and resources for resilience in childhood and adolescence. *Health and Place*, 28, pp. 1–13.

Chertow, M.R., 2007. "Uncovering" industrial symbiosis. *Journal of Industrial Ecology*, 11(1), pp. 11–30.

Chester, M.V. and Allenby, B., 2018. Toward adaptive infrastructure: Flexibility and agility in a non-stationarity age. *Sustainable and Resilient Infrastructure*, 4(4), pp. 173–191.

City of Amsterdam, Circle Economy and Copper8, 2017. Amsterdam circular: Evaluation and action perspectives, Report published by the City of Amsterdam.

City of Stockholm, 2015. Stockholm Royal Seaport: Leading the way towards a sustainable future, Sustainability Report, City of Stockholm 2015.

City of Stockholm, 2017. *Sustainable urban development programme 2017*. City of Stockholm.

City of Stockholm, 2018. *Stockholm Royal Seaport- sustainability report 2017*. Stockholm City Council.

Collado, S., Staats, H. and Corraliza, J.A., 2013. Experiencing nature in children's summer camps: Affective, cognitive and behavioural consequences. *Journal of Environmental Psychology*, 33, pp. 37–44.

Communication, 2016. Group interview with Stockholm City Council and Stockholm Royal Seaport Teams.

Corburn, J., 2009. Cities, climate change and urban heat island mitigation: Localising global environmental science. *Urban Studies*, 46(2), pp. 413–427.

Corvalán, C., Hales, S. and McMichael, A.J., 2005. *Millenium ecosystem assessment—Ecosystems and human well-being: Health synthesis* (p. 3). Geneva: WHO.

Coutard, O. and Rutherford, J., 2011. Energy transition and city-region planning: Understanding the spatial politics of systemic change Ag: 220. *Operations Research Management Science*, 51(3), p. 207.

Curtis, F., 2003. Eco-localism and sustainability. *Ecological Economics*, 46(1), pp. 83–102.

da Cruz, N.F. and Marques, R.C., 2012. Mixed companies and local governance: No man can serve two masters. *Public Administration*, 90, pp. 737–758. doi:10.1111/j.1467–9299.2011.02020.

Dadvand, P., Nieuwenhuijsen, M.J., Esnaola, M., Forns, J., Basagaña, X., Alvarez-Pedrerol, M., Rivas, I., López-Vicente, M., De Castro Pascual, M., Su, J., Jerrett, M., Querol, X. and Sunyer, J., 2015. Green spaces and cognitive development in primary school children. *Proceedings of the National Academy of Science*, 112, pp. 7937–7942.

Daily, G.C. and Ellison, K., 2012. *The new economy of nature: The quest to make conservation profitable*. Washington: Island Press.

Darby, S., 2006, The effectiveness of feedback on energy consumption. *A Review for DEFRA of the Literature on Metering, Billing and Direct Display*, 486, p. 26.

De Ceuvel Website. https://deceuvel.nl/en/about/sustainable-technology (accessed 24-04-20).

De Flander, K., 2015. Closed cycles-open city. In Craig Johnson, Noah Toly and Heike Schroeder (Eds.), *The Urban Climate Challenge: Rethinking the Role of Cities in the Global Climate Regime* (pp. 37–59). New York: Routledge.

De Groot, R.D., Fisher, B., Christie, M., Aronson, J., Braat, L., Haines-Young, R., Gowdy, J., Maltby, E., Neuville, A., Polasky, S. and Portela, R., 2010. Integrating the ecological and economic dimensions in biodiversity and ecosystem service valuation. In Kumar P. (Ed.), *The Economics of Ecosystems and Biodiversity (TEEB): Ecological and Economic Foundations* (pp. 9–40). New York: Routledge.

De Groot, R.S., Blignaut, J., Van der Ploeg, S., Aronson, J., Elmqvist, T. and Farley, J., 2013. Benefits of investing in ecosystem restoration. *Conservation Biology*, 27(6), pp. 1286–1293.

Demailly, K.È. and Darly, S., 2017. Urban agriculture on the move in Paris: The routes of temporary gardening in the neoliberal city. *ACME: An International Journal for Critical Geographies*, 16(2), pp. 332–361.

Dembski, S., 2013. Case study Amsterdam Buiksloterham, The Netherlands: The challenge of planning organic transformation. CONTEXT Report 2. AISSR programme group Urban Planning, Amsterdam.

Demuzere, M., Orru, K., Heidrich, O., Olazabal, E., Geneletti, D., Orru, H., Bhave, A.G., Mittal, N., Feliu, E. and Faehnle, M., 2014. Mitigating and adapting to climate change: Multi-functional and multi-scale assessment of green urban infrastructure. *Journal of Environmental Management*, 146, pp. 107–115.

Dooling, S., 2009. Ecological gentrification: A research agenda exploring justice in the city. *International Journal of Urban and Regional Research*, 33(3), pp. 621–639.

EC-European Commission, 2014. Towards a circular economy: A zero waste programme for Europe. *Brussels*, 2, p. 2014.

EC-European Commission, 2015. Closing the loop—An EU action plan for the circular economy. *Brussels*, 2.12.**2015** COM.

Ehrenfeld, J. and Chertow, M.R., 2002. Industrial symbiosis: The legacy of Kalundborg. In: Ayres, R.U., Ayres, L.W., Eds., *A Handbook of Industrial Ecology* (pp. 334–348). Cheltenham, UK: Edward Elgar.

Ellen Macarthur Foundation, 2019. Completing the picture: How the circular economy tackles climate change, Report published by EMF, Isle of Wight.

Ellen MacArthur Foundation, SUN, McKinsey Centre for Business and Environment, 2015. Growth within: A circular economy vision for a competitive Europe, Report published by Ellen MacArthur Foundation.

EU Waste Shipment Statistics, 2014. http://ec.europa.eu/trade/import-and-export-rules/export-from-eu/waste-shipment/index_en.htm (accessed 02-05-18).

Eunomia Research Consulting, 2016. A resourceful future–Expanding the UK economy, Report published by SUEZ.

Ferm, J. and Jones, E., 2016. Mixed-use 'regeneration 'of employment land in the post-industrial city: Challenges and realities in London. *European Planning Studies*, 24(10), pp. 1913–1936.

Fischer, C., 2008. Feedback on household electricity consumption: A tool for saving energy? *Energy Efficiency*, 1, pp. 79–104. doi:10.1007/s12053-008-9009-7.

Fischer-Kowalski, M. and Haberl, H., 1998. Sustainable development: Socio-economic metabolism and colonization of nature. *International Social Science Journal*, 50(158), pp. 573–587.

Folke, C., Jansson, A., Larsson, J. and Costanza, R., 1997. Ecosystem appropriation of cities. *Ambio*, 26(3), pp. 167–172.

Forest Research, 2010. Benefits of green infrastructure, Report by Forest Research. Forest Research, Farnham. http://www.forestry.gov.uk/pdf/urgp_benefits_of_green_infrastructure.pdf/$FILE/urgp_benefits_of_green_infrastructure.pdf

French National Statistics Agency (INSEE). https://www.insee.fr/en/accueil (accessed 24-04-20).

Furlong, K. and Bakker, K., 2010. The contradictions in alternative service deliver: Governance, business models, and sustainability in municipal water supply. *Environment Planning C*, 28, pp. 349–368. doi:10.1068/c09122.

Geddes, P., 1915. *Cities in evolution*. London: Williams and Norgate.

Gill, S., Handley, J., Ennos, A. and Pauleit, S., 2007. Adapting cities for climate change: The role of green infrastructure. *Built Environment*, 33(1), pp. 115–133.

Gómez-Baggethun, E. and Barton, D.N., 2013. Classifying and valuing ecosystem services for urban planning. *Ecological Economics*, 86, pp. 235–245.

Grazulevicience, R., Dedele, A., Danileviciute, A., Vencloviene, J., Grazulevicius, T., Andrusaityte, S., Uzdanaviciute, I. and Nieuwenhuijsen M.J., 2014. The influence of proximity to city parks on blood pressure in early pregnancy. *International Journal of Environmental Research and Public Health*, 11, pp. 2958–2972.

Greater London Authority, 2012. Queen Elizabeth Olympic Park Website, https://www.queenelizabetholympicpark.co.uk/the-park/plan-your-visit/park-map (accessed 24-04-20).

Greater London Authority, 2015a. Round demographic projections; and best foot forward: City limits—A resource flow and ecological footprint analysis of Greater London, published by City Hall, London.
Greater London Authority, 2015b. Bag it or bin it? Managing London's domestic food waste, published by City Hall, London.
Greater London Authority, 2017a. Responsible procurement policy, published by City Hall, London.
Greater London Authority, 2017b. Draft London plan, published by City Hall, London.
Greater London Authority, 2018a. London's environment strategy, published by City Hall, London.
Greater London Authority, 2018b. The London food strategy: Implementation plan 2018–2023, published by City Hall, London.
Gunderson, L., 2000. Ecological resilience in theory and application. *Annual Review of Ecology and Systematics*, 31(1), 425–439. doi:10.1146/annurev. ecolsys.31.1.425.
Hall, C.R. and Dickson, M.W., 2011. Economic, environmental, and health/well-being benefits associated with green industry products and services: A review. *Journal of Environmental Horticulture*, 29(2), pp. 96–103.
Hallegatte, S., Ranger, N., Mestre, O., Dumas, P., Corfee-Morlot, J., Herweijer, C. and Wood, R.M., 2011. Assessing climate change impacts, sea level rise and storm surge risk in port cities: A case study on Copenhagen. *Climatic Change*, 104, pp. 113–137.
Harlan, S.L. and Ruddell D.M., 2011. Climate change and health in cities: Impacts of heat and air pollution and potential co-benefits from mitigation and adaptation. *Current Opinion in Environmental Sustainability*, 3, pp. 126–134.
Harrap, C., 2019. World's largest urban farm to open—On a Paris rooftop, Guardian Newspaper 13-08-19 (accessed 24-04-20).
Herold, R. and Hertzog, C., 2015. *Data privacy for the smart grid*. Boca Raton, FL: CRC Press.
Hofstede, G., 2001. *Culture's consequences: Comparing values, behaviors, institutions and organizations across nations*. Beverly Hills, CA: Sage.
Institute for Urban Planning and Development of Paris Ile-de-France Region, 2016. *The environment in Ile-de-France in 2015*. Paris: IAU.
International Energy Agency, 2008. *World Energy Outlook 2008*. IEA.
Jackson, T., 2009. *Prosperity without growth: Economics for a finite planet*. London: Routledge.
Janssen, I. and Rosu, A., 2015. Undeveloped green space and free-time physical activity in 11 to 13-year-old children. *Journal of Behavioral Nutrition and Physical Activity*, 12, p. 26.
Jeppesen, B. and Solley, D., 1994. *Domestic greywater reuse: Overseas practice and its applicability to Australia*. Urban Water Research Association of Australia.
Kelz, C., Evans, G.W. and Röderer, K., 2013. The restorative effects of redesigning the schoolyard: A multi-methodological quasi-experimental study in rural Austrian middle schools. *Environment and Behavior*, 12(1), pp. 1–21.
Kennedy, C., Cuddihy, J. and Engel-yan, J., 2007. The changing metabolism of cities. *Journal of Industrial Ecology*, 11(2), pp. 43–59.
Kennedy, C., Pincetl, S. and Bunje, P., 2011. The study of urban metabolism and its applications to urban planning and design. *Environmental Pollution*, 159(8–9), pp. 1965–1973.
Khan, Z., Pervez, Z. and Ghafoor, A., 2014. Towards cloud based smart cities data security and privacy management. In: *Proceedings of the International Workshop on Smart City Clouds: Technologies, Systems and Applications in Conjunction with 7th IEEE/ACM Utility and Cloud Computing (UCC)* (pp. 806–811). London: IEEE.
Kocornik-Mina, A., McDermott, T.K., Michaels, G. and Rauch, F., 2020. Flooded cities. *American Economic Journal: Applied Economics*, 12(2), pp. 35–66.

Kovats, R.S. and Ebi, K.L., 2006. Heatwaves and public health in Europe. *European Journal of Public Health*, 16(6), pp. 592–599.

Krook, J., Svensson, N. and Eklund, M., 2012. Landfill mining: A critical review of two decades of research. *Waste Management*, 32(3), pp. 513–520.

Lacovidou, E. and Purnell, P., 2016. Mining the physical infrastructure: Opportunities, barriers and interventions in promoting structural components reuse. *Science of the Total Environment*, 557, pp. 791–807.

Lambeth Council. https://www.lambeth.gov.uk/better-fairer-lambeth/project/pop-brixton (accessed 24-04-20).

Latham, L., 2018. The rise of the 'meanwhile space': How empty properties are finding second lives, Guardian Newspaper 28-11-18 (accessed 24-04-20).

Laville, S. and Taylor, M., 2017. Stop exporting waste to China, Guardian Newspaper 29-7-17. https://www.theguardian.com/environment/2017/jun/29/stop-exporting-plastic-waste-to-china-to-boost-recycling-at-home-say-experts (accessed 02-05-18).

Lenhart, J., van Vliet, B. and Mol, A. 2015. New roles for local authorities in a time of climate change: The Rotterdam energy approach and planning as a case of urban symbiosis. *Journal of Cleaner Production*, 107, pp. 593–601. doi:10.1016/j.jclepro.2015.05.026.

Lewandowski, M., 2016. Designing the business models for circular economy—Towards the conceptual framework. *Sustainability*, 8(1), p. 43.

Liang, S. and Zhang, T., 2011. Urban metabolism in China achieving dematerialization and decarbonisation in Suzhou. *Journal of Industrial Ecology*, 15(3), 420–434.

London Legacy Development Corporation, 2014. Grass-root Interim Uses Project Guidelines, published by London Legacy Development Corporation, London.

London Legacy Development Corporation, 2017. Your park, our planet: Environmental sustainability report 2016–17, published by London Legacy Development Corporation, London.

London Sustainable Development Commission (2015) Employment and the circular economy: job creation through resource efficiency in London. Report by LSDC and GLA.

London Waste & Recycling Board, 2015. London the circular economy capital: Towards a circular economy–Context and opportunities, Report published by LWARB.

London Waste and Recycling Board, 2017. London's circular economy route map, Report published by LWARB.

Love Lambeth. https://love.lambeth.gov.uk/funding-boost-for-loughborough-farm/ (accessed 24-04-20).

Maas, J., Verheij, R.A., de Vries, S., Spreeuwenberg, P., Schellevis, F.G. and Groenewegen, P.P., 2009. Morbidity is related to a green living environment. *Journal of Epidemiology and Community Health*, 63(12), pp. 967–973.

Madanipour, A., 2018. Temporary use of space: Urban processes between flexibility, opportunity and precarity. *Urban Studies*, 55(5), pp. 1093–1110.

Marie de Paris, 2015a. The blueprint for the non-potable water network and uses, published by the Marie de Paris, Paris.

Marie de Paris, 2015b. Clichy- Batignolles, published by the Marie de Paris, Paris.

Marie de Paris, 2017. Circular economy plan Paris adopted 2017, published by Marie de Paris, Paris.

Marie de Paris, 2018. Paris climate action plan, published by the Marie de Paris, Paris.

Mazzucato, M., 2011. The entrepreneurial state: Debunking public vs private sector myths. New York: Public Affairs.

McDonough, W.B.M., 2002. *Cradle to cradle: Remaking the way we make things*. New York: North Point.

McKinsey Global Institute, 2016. *Urban world: The global consumers to watch*. New York: McKinsey & Co. https://www.mckinsey.com/;/media/mckinsey/featured%20insights/Urbanization/Urban%20world%20The%20global%20consumers%20to%20watch/Urban-World-Global-Consumers-Executive-summary.ashx (accessed 15-12-18).

McPhearson, T., Pickett, S.T., Grimm, N.B., Niemelä, J., Alberti, M., Elmqvist, T., Weber, C., Haase, D., Breuste, J. and Qureshi, S., 2016. Advancing urban ecology toward a science of cities. *BioScience*, 66(3), pp. 198–212.

McPherson, E.G., 1998. Structure and sustainability of Sacramento's urban forest. *Journal of Arboriculture*, 24, pp. 174–190.

McPherson, E.G., Nowak, D., Heisler, G., Grimmond, S., Souch, C., Grant, R. and Rowntree, R., 1997. Quantifying urban forest structure, function and value: The Chicago Urban Forest Climate Project. *Urban Ecosystems,* 1, pp. 49–61.

Meier, F. and Scherer, D., 2012. Spatial and temporal variability of urban tree canopy temperature during summer 2010 in Berlin, Germany. *Theory Applied Climatology*, 110, pp. 373–384.

Meltzer, G., 2000. Cohousing: Towards social and environmental sustainability, unpublished PhD Thesis, Department of Architecture, The University of Queensland, Brisbane.

Metabolic, 2013. Cleantech Playground: *A cleantech utility in Amsterdam North*, Version 2.2- February 11, 2013, InnovatieNetwerk Report Number: 13.2.312.

Metabolic and SGS Search, 2017. Roadmap to circular land issue: An introduction to circular building projects, published by the Amsterdam Metropolitan Authority.

Ministry for an Ecological and Solidarity Transition and Ministry of Economy and Finance, 2018. 50 Measures for 100% circular economy: A roadmap for a circular economy, Ministry for an Ecological and Solidary Transition and Ministry of Economy and Finance.

Mitchell, P., 2015, Employment and the circular economy—Job creation through resource efficiency in London, published by City Hall, London.

Mitchell, P. and James, K., 2015. Economic growth potential of more circular economies. Waste and Resources Action Programme (WRAP): Banbury, UK.

Moriguchi, Y., 2007. Material flow indicators to measure progress toward a sound material-cycle society. *Journal of Material Cycles and Waste Management*, 9(2), pp. 112–120.

Mumford, L., 1968. *The urban prospect*. New York: Harcourt Brace Jovanovich.

Municipal Council of Amsterdam, 2015. Sustainable Amsterdam: Agenda for renewable energy, clear air, a circular economy and a climate-resilient city, Report published by Municipal Council of Amsterdam.

Németh, J. and Langhorst, J., 2014. Rethinking urban transformation: Temporary uses for vacant land. *Cities*, 40, pp. 143–150.

Ness, D., Swift, J., Ranasinghe, D.C., Xing, K. and Soebarto, V., 2015. Smart steel: New paradigms for the reuse of steel enabled by digital tracking and modelling. *Journal of Cleaner Production*, 98, pp. 292–303.

Nguyen, B., 2003. Operation of dual drinking and non-potable water networks in Paris: Advantages and constraints. *Water Science and Technology: Water Supply*, 3(3), pp. 193–200.

Noordwaarts, 2009. *Bestemmingsplan "Buiksloterham"*. Amsterdam: Gemeente Amsterdam.

Nowak, D., 1994. Air pollution removal by Chicago's urban forest. In Gregory McPherson (Ed.), *Chicago's Urban Forest: Results for the Chicago's Urban Forest Climate Project* (pp. 63–81). Chicago: USDA Forest Service.

Nowak, D.J. and Dwyer, J.F., 2007. Understanding the benefits and costs of urban forest ecosystems. In *Urban and Community Forestry in the Northeast* (pp. 25–46). Springer, Dordrecht.

Okvat, H.A. and Zautra, A.J., 2011. Community gardening: A parsimonious path to individual, community, and environmental resilience. *American Journal of Community Psychology*, 47(3–4), pp. 374–387.

O'Neill, M.S., 2009. Preventing heat-related morbidity and mortality: New approaches in a changing climate. *Maturitas*, 64, pp. 98–103.

Orr, D., 1992. *Environmental literacy: Education as if the earth mattered*. Bristol: E. F. Schumacher Lectures.

Ortner, M.E., Knapp, J. and Bockreis, A., 2014. Landfill mining: Objectives and assessment challenges. *Proceedings of the Institution of Civil Engineers*, 167, p. 51. doi:10.1680/warm.13.00012.

Pandis, S., 2014. Industrial ecology for sustainable urban development—The case of Hammarby Sjöstad. PhD Thesis. KTH Stockholm: Sweden.

Pandis, S., Johanssen, S. and Brandt, N., 2013. The potential of the infrastructural system of Hammarby Sjöstad in Stockholm. *Energy Policy*, 59, pp. 716–726.

Parkins, A.E., 1930. Profiles of the retail business section of Nashville, Tenn., and their interpretation. *Annals of the Association of American Geographers*, 20(3), pp. 164–175.

Partnership Circular Economy, 2017. *Urban agenda for the EU—Partnership circular economy*. Amsterdam, The Netherlands: Partnership Circular Economy.

Peng, S., Piao, S., Ciais, P., Friedlingstein, P., Ottle, C., Bréon, F.-M., Nan, H., Zhou, L. and Myneni, R.B., 2012. Surface Urban heat island across 419 global big cities. *Environmental Science and Technology*, 46(2), pp. 696–703.

Perchard, E., 2016. French food waste law passes unanimously, published in Resource Magazine, 03-04-16 (accessed 24-04-20).

Pilsudski, T. and Koh, M., 2019. Reinventer Paris (Reinventing Paris) innovation as a key consideration for land sale sites, published by Centre for Liveable Cities, Singapore (accessed 24-04-20).

Pincetl, S., Bunje, P. and Holmes, T., 2012. An expanded urban metabolism method: Toward a systems approach for assessing urban energy processes and causes. *Landscape Urban Planning*, 107, pp. 193–202.

Port of Amsterdam, 2018. Amsterdam ready for the Bio-based Port of Amsterdam Economy, published by the Port of Amsterdam.

Ports of Stockholm, 2018. Annual report and sustainability report 2017, Ports of Stockholm, Stockholm, Sweden.

Radcliff, J., 2003. An overview of water recycling in Australia-results of a recent ATSE study. In: *Proceedings of Water Recycling Australia: Second National Conference*, Brisbane, 1–3 September 2003.

Ranhagen, U. and Frostell, B., 2014. Eco-cycle model 2.0 for Stockholm Royal Seaport City District: Final Report, City of Stockholm and KTH School of Architecture and the Built Environment.

Rauws, W. and De Roo, G., 2016. Adaptive planning: Generating conditions for urban adaptability. Lessons from Dutch organic development strategies. *Environment and Planning B: Planning and Design*, 43(6), pp. 1052–1074.

Regional Authority of Greater Paris, 2013. Regional master plan—Greater Paris region 2030 (SDRIF plan), published by Regional Authority of Greater Paris.

Richter, B.D., Abell, D., Bacha, E., Brauman, K., Calos, S., Cohn, A. and Siegfried, E., 2013. Tapped out: How can cities secure their water future? *Water Policy*, 15(3), pp. 335–363.

Rigolon, A. and Németh, J., 2018. "We're not in the business of housing:" Environmental gentrification and the nonprofitization of green infrastructure projects. *Cities*, 81, pp. 71–80.

Rink, D., Haase, A., Bernt, M. and Grossmannn, K., 2012. *Shrink smart: The governance of shrinkage within a European context*. Leipzig: Helmholtz Centre for Environmental Research.

Roelich, K., Knoeri, C., Steinberger, J., Varga, L., Blythee, P., Butler, D., Gupta, R., Harrison, G., Martini, C. Purnell, P., 2015. Towards resource-efficient and service-oriented integrated infrastructure operation. *Technology Forecasting Society Change*, 92, pp. 40–52. doi:10.1016/j.techfore.2014.11.008.

Rosales, N., 2017. How can an ecological perspective be used to enrich cities planning and management?. *urbe. Revista Brasileira de Gestão Urbana*, 9(2), pp. 314–326.

Rougé, N., 2015. Clichy-Batignolles: Where urban planning meets the climate. *Organisation for Economic Cooperation and Development. The OECD Observer*, 304, p. 53.

Roy, S., Byrne, J. and Pickering, C., 2012. A systematic quantitative review of urban tree benefits, costs, and assessment methods across cities in different climatic zones. *Urban Forestry & Urban Greening*, 11(4), pp. 351–363.

Rydin, Y. and Pennington, M., 2000a. Public participation and local environment planning: The collective action problem and the potential for social capital. *Local Environment*, 5(2), pp. 153–169.

Sassen, S., 2015. Who owns our cities—And why this urban takeover should concern us all, Guardian Newspaper 24-11-2015. https://www.theguardian.com/cities/2015/nov/24/who-owns-our-cities-and-why-this-urban-takeover-should-concern-us-all (accessed 02-05-18).

Schwartz, S.H., 2008. *Cultural value orientations: Nature and implications of national differences*. Jerusalum, Israel: The Hebrew University.

Science Education through Earth Observation website. https://seos-project.eu/landuse/landuse-c02-p24.html (accessed 23-04-20).

Scottish Government, 2016. Making things last: A circular economy strategy for Scotland, Report published by Scottish Government.

Seto, K., Güneralp, B. and Hutyra, L., 2012. Global forecasts of urban expansion to 2030 and direct impacts on biodiversity and carbon pools. *Proceedings of the National Academy Science USA*, 109, pp. 16083–16088. doi:10.1073/pnas.1211658109.

Shahrokni, H., Lazarevic, D. and Brandt, N., 2014. Smart urban metabolism: Toward a real-time understanding of the energy and material flows of city and its citizens. *Urban Technology*, 22, pp. 65–86.

Simmonds, P.L., 1862. *Waste products and undeveloped substances: Or, hints for enterprise in neglected fields*. R. Hardwicke.

Skelhorn, C., Lindley, S. and Levermore, G., 2014. The impact of vegetation types on air and surface temperatures in a temperate city: A fine scale assessment in Manchester, UK. *Landscape and Urban Planning*, 121, pp. 129–140.

Smith, A., 2007. Emerging in between: The multi-level governance of renewable energy in the English regions. *Energy Policy*, 35, pp. 6266–6280. doi:10.1016/j.enpol.2007.07.023.

Stigsdotter, U.K., Ekholm, O., Schipperijn, J., Toftager, M., Kamper-Jørgensen, F. and Randrup, T.B., 2010. Health promoting outdoor environments—Associations between green space, and health, health-related quality of life and stress based on a Danish national representative survey. *Scandinavian Journal of Public Health*, 38(4), pp. 411–417.

Stockholm City Council. 2015. Stockholm Royal Seaport: Leading the way towards a sustainable future, Sustainability Report 2015

Stockholm City Council Website. http://www.hammarbysjostad.se/ (accessed 24-04-20).

Swickard, T.J., 2008. Regulatory incentives to promote private sector brownfield remediation and reuse. *Soil and Sediment Contamination*, 17, pp. 121–136. doi:10.1080/15320380701870393.

Tabuchi, J.P., Tassin, B. and Blatrix, C. Greater Paris water and global change. Working Paper. http://eaumega.org/wp-content/uploads/2016/05/EN-Paris-Monograph.pdf (accessed 24-04-20).

Taylor Wimpey. https://www.taylorwimpey.co.uk (accessed 24-04-20).

Teisman, G. and Gerrits, L., 2014. The emergence of complexity in the art and science of governance. *Complexity, Governance & Networks*, 1(1), pp. 17–28.

ten Brink, P., Mutafoglu, K., Schweitzer, J.-P., Underwood, E., Tucker, G., Russi, D., Howe, M., Maréchal, A., Olmeda, C., Pantzar, M. and Kettunen, M., 2017. *Natura 2000 and Jobs: Scoping study—Executive summary*. Brussels.

Thames Water, 2019. *Revised draft water resources management plan 2019*. London: Thames Water.

The Dutch Ministry of Infrastructure and the Environment and the Ministry of Economic Affairs, 2016. A circular economy in the Netherlands by 2050 Government-wide programme for a circular economy, report published the Government of the Netherlands.

Townsend, A.M., 2013. Smart cities: Big data, civic hackers, and the quest for a New Utopia. New York: WW Norton & Company.

Trevors, J.T. and Saier, M.H., 2010. The nature connection. *Water, Air, and Soil Pollution*, 205(1), pp. 85–86.

Ueno, T., Sano, F., Saeki, O. and Tsuji, K., 2006. Effectiveness of an energy-consumption information system on energy savings in residential houses based on monitored data. *Applied Energy*, 83, pp. 166–183.

UNEP, 2006. *Circular economy: An alternative for economic development*. Paris: UNEP DTIE UNEP.

United Nations, 2014. *World urbanisation prospects report*. New York: United Nations.

United Nations, D. o. E. a. S. A., Population Division., 2016. The World's Cities in 2016. http://www.un.org/en/development/desa/population/publications/pdf/urbanization/the_worlds_cities_in_2016_data_booklet.pdf (accessed 02-05-18).

United Nations, 2017. Report of the special rapporteur on adequate housing as a component of the right to an adequate standard of living and on the right to non-discrimination in this context A/HRC/34/51.

Unruh, G.C., 2000. Understanding carbon lock-in. *Energy Policy*, 28, pp. 817–830. doi:10.1016/S0301–4215(00)00070-7.

Van der Hoek, J.P., de Fooij, H. and Struker, A., 2016. Wastewater as a resource: Strategies to recover resources from Amsterdam's wastewater. *Resources, Conservation and Recycling*, 113, pp. 53–64.

Van Renterghem, T., Forssén, J., Attenborough, K., Philippe, J., Defrance, J., Hornikx, M. and Kang J., 2015. Using natural means to reduce surface transport noise during propagation outdoors. *Applied Acoustics*, 92, pp. 86–101.

Velis, C., 2015. Circular economy and global secondary material supply chains. *Waste Managemnet Research*, 33, pp. 389–391. doi:10.1177/0734242X15587641.

Voicu, I. and Been, V., 2008. The effect of community gardens on neighboring property values. *Real Estate Economics*, 36(2), pp. 241–283.

Voulvoulis, N., 2015. *The circular revolution*. Veolia: London.

Ward Thompson, C., Aspinall, P. and Montarzino, A., 2008. The childhood factor. Adult visits to green places and the significance of childhood experiences. *Environment and Behavior*, 40(1), pp. 111–143.

WHO, 2011. Burden of disease from environmental noise, Quantification of healthy life years lost in Europe. WHO regional office for Europe, Denmark.

Wilcox, J., Nasirib, F., Bell, S. and Rahaman, S., 2016. Urban water reuse: A triple bottom line assessment framework and review. *Sustainable Cities and Society*, 27, pp. 448–456. doi:10.1016/j.scs.2016.06.021.

Williams, J., 2005. Sun, surf and sustainable housing—Cohousing, the Californian experience. *International Planning Studies*, 10(2), pp. 145–177.

Williams, J., 2012. *Zero-carbon homes: A road map*. Routledge.

Williams, J., 2013. The role of planning in delivering low-carbon urban infrastructure. *Environment and Planning B: Planning and Design*, 40(4), pp. 683–706.

Williams, J., 2016. Can low carbon city experiments transform the development regime? *Futures*, 77, pp. 80–96.

Williams, J., 2017. Lost in translation: Translating low carbon experiments into new spatial contexts viewed through the mobile-transitions lens. *Journal of Cleaner Production*, 169, pp. 191–203.

Williams, J., 2019a. Circular cities. *Urban Studies*, *56*(13), pp. 2746-2762.

Williams, J., 2019b. Circular cities: Challenges to implementing looping actions. *Sustainability*, 11(2), p. 423.

Williams, J., 2019c. The circular regeneration of a Seaport. *Sustainability*, 11(12), p. 3424.

Woetzel, J., Remes, J., Coles, K. and Krishnan, M., 2016. *Urban world: Meeting the demographic challenge in cities*. McKinsey Global institute.

Wolman, A., 1965. The metabolism of cities. *Scientific American*, 213(3), pp. 179–190.

World Health Organization, 2018. *WHO global ambient air quality database (update 2018)*. Geneva, Switzerland: World Health Organization.

World Wide Fund, 2016. Living Planet Report 2016, published by Panada House, London.

WRAP, 2015. Economic growth potential of more circular economies.

Zhang, Y., 2013. Urban metabolism: A review of research methodologies. *Environmental Pollution*, 178, pp. 463–473.

Zoulia, I., Santamouris, M. and Dimoudi, A., 2009. Monitoring the effect of urban green areas on the heat island in Athens. *Environmental Monitoring Assessment*, 156, pp. 275–292.

INDEX

Note: **Bold** page numbers refer to tables, *Italic* page numbers refer to figures and page number followed by "n" refer to end notes.

For Product Safety Concerns and Information please contact our EU
representative GPSR@taylorandfrancis.com
Taylor & Francis Verlag GmbH, Kaufingerstraße 24, 80331 München, Germany

www.ingramcontent.com/pod-product-compliance
Ingram Content Group UK Ltd.
Pitfield, Milton Keynes, MK11 3LW, UK
UKHW021307260926
475702UK00048B/799
* 9 7 8 1 1 3 8 5 9 1 1 4 1 *